P9-DNR-621

The Role
of International
Companies
in Latin American
Integration

**A Supplementary Paper of the
Committee for Economic Development**

The Role
of International
Companies
in Latin American
Integration

Autos and Petrochemicals

Jack N. Behrman
School of Business Administration
University of North Carolina

Published for the Committee for Economic Development

Lexington Books
D.C. Heath and Company
Lexington, Massachusetts
Toronto   London

COMMITTEE FOR ECONOMIC DEVELOPMENT
477 Madison Avenue, New York, N.Y. 10022
Library of Congress Catalog Card Number 79-183711
International Standard Book Number, paperback edition 0-87186-235-2

Table of Contents

List of Tables and Figures

Table

Figure

Foreword

Various initiatives of Latin American countries to move toward regional integration as a means of advancing their economic development have thus far made relatively little progress. The most ambitious of these, the Latin American Free Trade Association (LAFTA) launched in 1960, has sought to achieve regional integration through tariff reductions across the range of goods traded among these nations. The tariff liberalization efforts of LAFTA have come to an almost complete standstill. Latin American geography, inequality of development levels in the different countries, and strong nationalism, together with the complexity of balancing national advantages, are among the reasons for this disappointing showing. It now seems most unlikely that the region will have made substantial progress through across-the-board tariff reductions to achieve even the foundation of a Latin American Common Market by 1985, which was the target date announced for completion of the Market by the American presidents at the Punta del Este conference in 1967.

Other efforts to promote Latin American integration have been carried on at levels that are lower keyed and less ambitious than LAFTA, both as to the number of countries involved and the industries covered. One of the more promising of these efforts relates to sectoral trade arrangements. Known under the LAFTA pact as complementation agreements, these arrangements are aimed at achieving free trade in specific industrial sectors and allocating the required production facilities among the countries that enter into the agreements. While by no means easy to negotiate or implement, sectoral agreements focus the problems in a sharp and practical manner and limit to a narrow range of products the need to balance the benefits and losses of the different countries, thereby simplifying the process of intergovernmental bargaining.

Various Latin American nations, particularly the countries in the Andean Pact subregion, have found complementation agreements sufficiently attractive to justify extensive experimentation with them. The following study notes that at the time of its publication nine such agreements were operating, seven had been signed but were not yet in operation, and another sixteen were in various stages of negotiation. The sectors involved ranged from pharmaceuticals and domestic appliances to communications and office equipment. As yet, the actual volume of trade encompassed by these agreements is not very great, and it would be premature to forecast the degree of success likely to be achieved. However, the complementation agreements command attention as one of the few methods—if not indeed the only one—now showing discernible movement toward regional economic integration.

International companies with operations in several Latin American countries are in an exceptionally favorable position to form or participate in complementation agreements since they not only have the necessary contacts but also often hold the technological, managerial, and marketing keys to the success of the industry. A deterrent for some of these companies is that they have made investments in two or more countries as a result of promised protection by each government from competition within its jurisdiction. As a result, in some industries there is duplication and over-capacity of facilities for the region without much reference to what might be an efficient regional ordering of industrial activity. A different but per-haps even more important problem is that if major international companies were to be integrated regionally their influence might become greater than would be acceptable to at least some of the governments. This objection might be met by complementation agreements that linked international companies with nationalized firms and privately owned domestic businesses and supplier companies—at the risk, however, of encountering difficulties in achieving efficient management.

This study analyzes the possibilities and problems of making and operat-ing complementation agreements in Latin America in two important indus-trial sectors, automotive and petrochemical. These are not entirely new fields for research. The chemical industry has been studied by LAFTA and the United Nations Economic Commission for Latin America (ECLA), both of which have recently published substantial reports on the structure and costs of the industry. The automobile industry has been the subject of extensive examination by LAFTA, by several individual authors, and by a European company with affiliates in Latin America. These studies, however, did not include any considerable analysis of the role of the international companies in the process of regional integration. This vitally important aspect is the particular emphasis of the present study.

The author of the study, Jack N. Behrman of the School of Business Administration at the University of North Carolina, has long been a student of Latin American economic integration. In this field, Professor Behrman guided and advised in the preparation of two doctoral dissertations, one on the petrochemical industry by Thomas Goho, and the other on the auto-motive industry by James Fox. These two studies, which appear as ap-pendixes to this volume in condensed form, provided a data base and tentative analyses of some of the problems discussed in the main text. In assessing the benefits of and the obstacles to regional integration in chem-icals and autos in Latin America, the author and his colleagues used the available studies but went further, soliciting through interviews the com-ments and observations of officials of companies and governments in the region during the summer of 1969, with a brief follow-up in the summer of the following year.

This study is a continuation of the Committee for Economic Development's interest in Latin America going back a decade. In this period, CED has issued several policy statements and other studies dealing with various aspects of economic development and trade in the low-income countries. Two of these statements and two research publications were devoted specifically to Latin America.* The CED is pleased to add the present volume as a further contribution to the understanding of the problems of economic development and the processes by which it may be achieved.

Roy Blough, *Director*
CED *International Economic Studies*
Columbia University

* The Statements on National Policy by the CED Research and Policy Committee are *Cooperation for Progress in Latin America* (1961) and *Economic Development of Central America* (1964). The CED Supplementary Papers are *Economic Development Issues: Latin America* (1967), and *Regional Integration and the Trade of Latin America* (1968).

Acknowledgments

I would like to express appreciation for the assistance given me in the preparation of this study by Thomas Goho and James Fox, whose dissertations on the automotive and petrochemical industries are condensed in the Appendixes. Since I have not always agreed with their assessments and have included my own selection of problems and their analysis, the responsibility for any errors of interpretation found in the main body of the text is entirely mine.

I am grateful to the Committee for Economic Development for the opportunity to undertake this study through its sponsorship of the research, and I would also like to thank members of the Editorial Board; Roy Blough, Director of CED International Economic Studies; and members of the CED staff for their highly useful editorial suggestions.

Jack N. Behrman

A CED SUPPLEMENTARY PAPER

Supplementary Paper Number 35 is issued by the Research and Policy Committee of the Committee for Economic Development in conformity with the CED Bylaws (Art. V, Sec. 6), which authorize the publication of a manuscript as a Supplementary Paper if:

a) It is recommended for publication by the Project Director of a sub-committee because in his opinion, it "constitutes an important contribution to the understanding of a problem on which research has been initiated by the Research and Policy Committee" and,

b) It is approved for publication by a majority of an Editorial Board on the ground that it presents "an analysis which is a significant contribution to the understanding of the problem in question."

This Supplementary Paper relates to the following Statements on National Policy issued by the CED Research and Policy Committee: *Cooperation for Progress in Latin America* (1961), *Economic Development of Central America* (1964), and *How Low Income Countries Can Advance Their Own Growth* (1966).

The members of the Editorial Board authorizing publication of this Supplementary Paper are:

This paper has also been read by the Research Advisory Board, the members of which under CED Bylaws may submit memoranda of comment, reservation, or dissent.

While publication of this Supplementary Paper is authorized by CED's Bylaws, except as noted above its contents have not been approved, disapproved, or acted upon by the Committee for Economic Development, the Board of Trustees, the Research and Policy Committee, the Research Advisory Board, the Research Staff, or any member of any board or committee, or any officer of the Committee for Economic Development.

**The Role
of International
Companies
in Latin American
Integration**

1 Introduction and Summary

The ideal state of international economic integration, in the minds of many economists, is that arising from free trade. But it is unlikely that we can reach that utopian state,* for free trade is not wholly acceptable to any nation, including the United States, because the results are considered inequitable from the standpoint of national interests. Still, the efficiencies of specialization under freer trade are highly desirable, and all countries want greater exports to help cut the costs of domestic industrial production.

The most useful approach to the further freeing of trade and making production more efficient seems to be that proposed by Eric Wyndham White in the last months of his tenure as director-general of the General Agreement on Tariffs and Trade (GATT). He argued that the next stage of negotiations over trade barriers should take each industrial sector separately —in whatever bite size would permit agreement among nations. Only by focusing on the special situations arising in the sectors of, say, paper and pulp, aluminum, iron and steel, electronics, automobiles, and chemicals, would the problems become tractable. The present study starts from the position that increasingly trade will be a result of foreign investment and the international location of production. A sectoral approach will be needed to weld the patterns of investment and trade so as to achieve both efficiency and equity—"efficiency" in the market sense of least-cost production and "equity" in the sense of an acceptable distribution among countries of the benefits of industrial advance. For national governments, equity is denominated in terms of the locus of production and of impacts on balances of payments, employment, and technology transfers. The mere *re*distribution of income is not an appropriate technique to achieve equity, which must include not only direct revenue benefits but also provide for direct *participation* in the process of industrial growth.

The lack of means of guaranteeing equity (i.e., reciprocity or a balancing of benefits and burdens) among countries has slowed the progress of Latin America toward economic integration under the Latin American Free Trade Association (LAFTA) agreements. This goal is still strongly supported by officials in both government and business, and ways of stimulating new moves have been sought. But a major obstacle in some sectors appears to be the existence of affiliates of international companies, which would be likely

---

* Like all utopias, it is utopian only in the minds of those proposing it *and* because it selects *one* criterion to maximize.

1

to benefit from further regional integration; in the Declaration of the American Presidents (issued at Punta del Este, Uruguay, in 1967) it was asserted that integration will not be accomplished in ways which singularly benefit foreign-owned companies. Yet, integration remains a goal, and the international companies can make a contribution to that goal.

Obviously, other economic and political goals are sought besides integration, such as greater national prestige through industrial and technological advance, reduced unemployment, and more consumer goods for the lower-income groups in the country. These goals, if pursued by themselves, would likely dictate quite different attitudes toward sectoral industrial growth. And some of them may be in conflict with the goal of integration, which would probably result in higher-cost goods in some sectors than would imports from a world supplier and might require technologies that did not employ the largest number of workers. Building a prestigious industrial base also could be contrary to the structure of industry under integration.

Because individual countries show a propensity at times for one or another of these alternative goals over the goal of integration, no attempt is made here to prescribe which goal should be pursued. Assuming that as stated a major objective of governmental officials is economic integration, we have sought to examine the prospects in two industries, given the existence of the international companies in each.

Specific proposals have been made for sectoral integration in automobiles and in petrochemicals; one agreement has been signed in petrochemicals, but it covers only a few of the products and only a few Latin American countries are signatories. Successful regional or subregional integration in either industry waits on further moves, which appear difficult for Latin America to make.

Some of the difficulties surfaced in the early efforts to move toward integration through reduction of national barriers to trade. Three different mechanisms were provided in the early agreements on Latin American integration for the reduction of trade barriers among the members: the national lists, the common lists, and complementation agreements. The national lists were to include items on which members were to reduce their over-all duties by 8 per cent each year for twelve years, thus achieving free trade in gradual steps by the end of the period. The reductions related only to items already traded; moreover, a concession on any item could be removed at any time and replaced by equal reductions on another item. The common lists were to include items on which *all* members would extend free-trade treatment only at the *end* of the transitional twelve-year period; 25 per cent of the dutiable items were to be placed on the list each three years. The complementation agreements were to be used to extend free trade or preferential treatment at any time within an industrial sector, not waiting for the products to be placed on either of the lists.

In the earlier years of LAFTA, there was little difficulty with either the national or common lists. Under the national list, countries reduced duties on items they wanted to import; they placed items on the common list that were not traded among themselves. Negotiations began to stagnate, however, as they began to get to items that were significant in present trade (in which they had a local production interest) or that would be important in future trade. In the past few years, it has become evident that further moves under these lists are unlikely to be taken.

The reasons for this impasse have to do with the different assessments by the member countries of the impacts of free trade upon their economies. The less-advanced countries feel they are being asked to give up their prospects for industrialization in favor of those countries already advanced industrially. The more-advanced countries feel they are being asked to make all the sacrifices for integration; some of the advanced countries (e.g., Brazil) even feel they are bearing a burden for other advanced countries. The crux of the difficulty is that no country is willing to accept the conditions of production and trade occurring under free trade.

As has occurred elsewhere, free trade would give rise to mergers in companies, with some occurring transnationally within Latin America; ownership of industry would shift, as would the patterns of trade and employment. It is not at all evident that, under free trade, industry is spread widely or evenly over a region that is integrated economically (*vide* the experience within Italy, France, England, Germany, and the United States). Rather, industry tends to concentrate where certain advantages are found—including that of an early start.

Since industry location under present conditions is seen as fortuitous, members of an integration group are not eager to accept such a pattern for the future. Even if there were no established pattern, it is unlikely that members would accept a free-market determination of the location of industry, for there is nothing in the operation of a free market that automatically provides an equitable distribution of benefits among members arising from industrial diversification. Only with complete mobility of *all* factors could the system come close to achieving equity; and, of course, mere reduction of duties does not provide that mobility.

In the absence of a free market, the location of industry will be determined or be affected strongly by national governments or by those who control private industry. Neither source is a satisfactory decision maker for other nations. Each government is more interested in the welfare of its own national economy than in the welfare of the whole. Furthermore, private industrial companies are ethnocentric in their interests; and in Latin America, many of those who control such companies are not even Latin Americans. The removal of national control through extension of free trade, therefore, would transfer much of the control outside of Latin

America. The criteria for the decisions by such foreign control centers would undoubtedly not be those of the various member nations, and the results would not be acceptable. Even if the decisions turned out to be identical to those that would have been taken by the host government, the government could not accept it readily since it was taken by a foreign entity. (As one sage has said, "God made a great mistake when he made foreigners.") It is not immaterial to members of a nation *who* makes the decision affecting their welfare.

But there is another reason for the stagnation of the recent effort to reduce duties. It is simply that the balancing of benefits is too complex. Since members do not accept the assumption that free trade will benefit all, each tries to make certain that the concessions it gains will provide reciprocal benefits. To achieve this over a large range of concessions is so complex a task that negotiations can break down. A more manageable range of decisions is made available by the complementation agreement (or sectoral) arrangement provided for under the Treaty of Montevideo in 1960, which brought LAFTA into being. It was later modified so as to permit only those countries that are interested to join a given arrangement, removing the requirement that they extend most-favored-nation treatment to nonsignatories. Even these arrangements have not been easy to negotiate, for the removal of barriers gives an advantage to the already advanced countries; the location of future industry tends to be set in the pattern of existing industry, and since the duty reductions are supposed to be irrevocable (in contrast to those under the national lists) and immediate (rather than delayed as under the common list), countries are wary of entering into such an arrangement *unless* the distribution of industrial growth is also provided for in advance.

Finally, governments have delayed signing complementation agreements in order to gain the time to build up their own industry in preparation for future negotiations.

In order to gain *something* from the potentials of a wider market, governments have entered into temporary complementation agreements; these are called agreements on surpluses and deficits (*acuerdos de excedentes y faltantes*). They provide for the matching of shortages and surpluses in a given product among two or more countries to relieve overcapacity in one country and permit imports into another country that intends later to build capacity in the same line. The temporary nature of the agreements permits pursuit of the national industrial objective in the longer run. The existence of such agreements indicates the strength of the national orientation to industrial policy in Latin America.

From the problems met in trying to use the traditional means of economic integration, it has become clear that new techniques, solutions, or mechanisms are necessary to foster integration in Latin America. This conclusion has been reached by many Latin American officials—Raúl Prebisch, Felipe Herrera, Fernando Lara Bustamente, Carlos Sanz de Santamaria, and Julio Alberto Lagos, to name a few. Their responses to the need differ in breadth and emphasis, but all have pointed to the need for new efforts at negotiating complementation agreements.

The form and scope of such agreements is critical to their successful negotiation and implementation. To date, the ones negotiated have been narrow in scope, and only a few countries have signed. Nothing on the order of an industrywide agreement for Latin America has been attempted. Among the sixteen signed to date, only three provide for means of allocating industrial plants; the others are mere tariff-reduction agreements. Given the criterion of reciprocity in benefits, and the unwillingness to accept the benefits as determined under free trade, what is needed paramountly in such agreements is a means of providing for "fair shares" in production as well as trade. Otherwise, some countries could be left in an economic backwater.

At the same time, the governments know that it is necessary to expand regional exports to the United States, Europe, and elsewhere; therefore, the new industrial structure must eventually provide worldwide competitiveness. The crux of the problem is to find an appropriate trade-off between efficiency (in an economic sense) and equity (in the sense of division of benefits) that still permits worldwide competitiveness. In achieving equity, it may turn out that the costs of production in the industry are higher than those in the international market, though they should be lower than those under the separate national industries. Some means would have to be found of offsetting the higher costs to be able to export—at least until costs could be brought down through economies of scale.

Fortunately for national ambitions, the theory of comparative advantage—which has been used to show that a given activity should be located in one spot and not another—is not so rigorous that governments have felt that they had to follow it in determining the location of industrial activity, regardless of the cost to other national objectives. Rather it is feasible under present conditions of industrial production, to place industrial activity wherever there are adequate materials, skills, and capital; and it is economically feasible to import or acquire all three. Given an adequate market to serve, there are few industrial activities that cannot be economically located nearly anywhere in the world—"economically" meaning not necessarily with equal efficiency but within an acceptable range of return of productive factors and at prices reasonably competitive. The problem, then, is to develop the market and to structure the industry so as to serve the economy both efficiently and equitably.

Since each country within an integrating region is very conscious of the fact that by contributing its market it expands the total market, making possible more efficient operation, each wants to be assured of the benefits it can expect to derive from that market expansion before it will be willing to accept a complementation agreement for the industry. The agreement must meet certain criteria for sharing benefits even if this is at the expense of maximum efficiency. The trade-offs may be costly in efficiency lost, or not, depending partly on whether or not significant segments of the industry already are in being in the country and region. The existence of going concerns in an industry poses a very real obstacle to complementation agreements. It is costly to dismantle operating plants or even to redirect their production, and few governments of countries with plants have been willing to agree to such shifts. Perhaps even worse, since the existence of plants in the country improves its bargaining position for sharing benefits, governments often seek to delay negotiation of complementation agreements until substantial facilities have first been built in the country. As a result, the duplication of facilities is promoted and agreement to make the industry regionally complementary is discouraged, if not precluded.

The examination of the automotive and petrochemical industries in Latin America demonstrates that it is feasible to gain both efficiency and equity through integration of these sectors either over the whole area or within a subregional market of the larger countries. The potential gains from integration are greater when companies and countries are more competitive than complementary in their existing production, and this is the case with most of the national production in these two industries in Latin America.

Major factors needed for efficiency exist within the full LAFTA regional market: demand sufficient to support substantial levels of production, sufficient economies of scale at these levels, and additional nonscale reductions in cost. In automobiles, competitive efficiency might leave something to be desired compared to world prices; but in petrochemicals, the differential in efficiency remaining after full regional integration would be practically nil for a number of important products.

The study suggests that, under certain assumptions, equitable benefits can be gained through full integration and the allocation of production among several of the countries; expansion of existing facilities could be redirected without seriously altering the efficiency of the operations. These assumptions exclude transport costs, the problem of exchange rate instability, and foreign-government interference in rationalization plans. There is little doubt that transport costs will alter significantly the movement of petrochemicals, for example, just as special facilities for movement of completed vehicles can alter the feasibility of automobile export and import. These three factors have been omitted here simply to permit the analysis to be less complex. This study is therefore not predictive of the structure of industry that might obtain under complementation agreements. Rather, it is directed toward showing the types of analyses needed and the ways in which the international companies might be used effectively within any integration scheme covering either the entire region or a subregion.

Comparable results can be obtained—and probably achieved more easily politically—through integration of the three largest countries only: Mexico, Brazil, and Argentina. The remaining cost differentials compared to world prices would be small. And the shifts in population to achieve equity would be easier; without such shifts it is clear that major benefits would accrue to one or two of the countries compared to the third.

The Andean subregion, however, is not in such a favorable position to achieve both efficiency and equitable sharing of benefits. Even if all countries of the subregion combined their demand, the market would remain too small to support internationally competitive operations in automobiles. It might be possible to become efficient in some petrochemicals, but *only* if one country alone were the producer, leaving the others as nonparticipants in that part of the industry. To achieve equity among members would raise the costs to such a level that there would be no way to enter the world market, and it is doubtful that the members would support the facilities. Moreover, if subregional integration proceeded, it would create costly hostages to full regional integration later, for the facilities created would not necessarily be those that would be allocated to Andean countries under full regional arrangements with larger production runs.

Though it is clear that gains arise from regional or subregional integration in both industries (and substantial gains arise in the case of the full region or the three major countries), governments are unlikely to move quickly to claim these gains. They have been trying, for the past decade or so, to obtain the benefits of industrialization for their national economies. They will forego these benefits only if their *share* in the regional benefits is larger than what they anticipate will be the benefits of continued national industrial development.

Given the differences in potential gains from the different approaches to industrial policy—national, subregional, and regional—there are several potential trade-offs available to Latin American countries. For example, under full regional integration, Mexico's petrochemical industry might be less important than Colombia's but still more beneficial to Mexico than under a national policy. Were there to be integration of the three biggest countries, Mexico's petrochemical industry would dominate the others. Rather than face such dominance, Brazil and Argentina might prefer full regional integration; they would maintain a better position as compared to Mexico, both relatively and absolutely. But in each case the national governments would lose control. To date, the major governments (as well as Ecuador, Colombia, and Peru, not to mention Bolivia), have sought to expand their national petrochemical complexes without regard to what the others are doing.

The benefits under nationally oriented industrial policies have been an increase in employment, acquisition of new technology, higher wages, new management skills, and new products, with some small exports. What has not been achieved is a sizable expansion of the home market or substantial exports to the world market, because of high prices and a lack of competitiveness. Governments adopted policies of inducing import substitution and thereby induced foreign investment in their countries. These high-cost facilities cannot easily (if at all) become export-oriented. The means used to induce investment and the continued national orientation of industrial policy has meant that the facilities have been bound to the national market. National facilities have even duplicated those elsewhere in Latin America, leaving little opportunity for intraregional exports.

The entrance of foreign companies meant that potentially important decisions would be made by foreigners. To reduce this effect, some governments have required that the auto companies obtain their supplies of accessories and major parts from local suppliers, often with a further requirement that the suppliers be locally owned. In petrochemicals, some of the governments own the production of intermediates, thus extending their ability to influence the production of final products.

Despite these constraints, the international companies dominate both industries in the final-product stage. These companies are the means of keeping the industries current and competitive to the degree they are. For the industries to become competitive worldwide, the international companies will have to be an integral part of the process of integration. It is only through these companies that Latin America can obtain the continued infusions or skills, technology, new products, management, and (probably) capital needed to keep pace with the advancement in the rest of the world— if this is the objective. Other routes to obtaining the same skills, proprietary rights, and assistance have been urged, but they are neither so certain nor so readily used that Latin America could be sure of not lagging behind. Present national industrial development in the industries could not have occurred without the international companies, and future growth is unlikely to be adequate without their participation. The terms of that participation are another question.

Regional integration in both automobiles and chemicals will have to be on the shoulders of the international companies, if the objective of world market penetration is to be achieved. In turn, these companies would be tied tightly to local suppliers—private companies in automobile and public enterprises in petrochemicals. Despite the ties to local suppliers, the result would be that some of the major benefits would accrue to or be under the control of international companies. This fact makes integration all the more difficult, for many officials have echoed the Declaration of American Presidents that integration should not occur for the benefit of foreign enterprises.

The existence of local suppliers and state enterprises in the two industries does not make integration any easier. On the contrary, they are themselves serious obstacles. The state enterprises in petrochemicals produce quite similar products, and integration of these companies would require a type of control similar to the European Coal and Steel Community (ECSC), with intergovernmental control of production, sales, prices and investments. The local auto-supply companies would have to shift out of some lines (meaning some would go out of business or retool completely) in favor of competitors in other countries. The political process makes this type of rationalization difficult. It would be easier to achieve integration if suppliers to both automobile and petrochemicals producers were vertically integrated with the international companies, for they could then be "commanded" by these companies under an agreed-upon integration scheme. This would not remove the tough decisions on equity and participation; it would merely make the mechanics of adjustment easier by avoiding the domestic political pressures.

The necessity to rationalize the locally owned suppliers means that some *national* control over the industry would be lost under the integration plan. What would be gained in its stead would be greater efficiency for the region, lower costs, expanded exports, continued inflows of technology, new products, and higher wages. These gains cannot be obtained at the national level by any one country in Latin America; the national market, present and future, is too small. Even the projected rapid expansions in Brazil are not likely to be enough to keep pace with the economies of scale that will be achieved elsewhere in the world in major automotive and petrochemical centers. It can be anticipated that the scale of production that is "economic" is likely to increase substantially over present levels; national growth in any Latin American country is not likely to be sufficient to reach these higher levels. But with integration of the entire region or the subregion of the three major countries, the growth is projected to be sufficient to reach and maintain worldwide competitiveness.

To achieve an integrated industrial structure, a new multinational mechanism will be required to implement complementation agreements. This could provide for the equitable as well as efficient (competitive) allocation of production among the members. Such a new mechanism is needed to make the trade-offs. A level of efficiency might be gained that is competitive worldwide without such agreements, but the benefits would accrue mainly to a few countries. The question is that of how much efficiency will be lost (if any) as equity is gained. The answer lies, first, in what criteria of equity is accepted; second, in what means are adopted in fitting the three production groups (local private companies, state enterprises, and the international companies) into these criteria; and third, in what control is exercised over implementation of the accord.

The objectives of integration and an equitable balance of the benefits can be adequately served by concentrating on the impacts of the agreement on cost reduction, balance of payments, balance of technology, and balance of employment among members. These are the areas over which governments show most concern and in which they seek to measure economic benefits. A precise balance of effects is not possible in each of these areas, but an acceptable balance is achievable—given political will. Since adjustments will have to be made to compensate for errors of prediction, there is a necessity for a control or decision-making mechanism that operates periodically to reassess and make adjustments, or at least that makes recommendations for adjustment which the members can approve or not.

This entity can be either an intergovernmental body, or a new Latin American multinational enterprise, or the international companies, with the latter two under governmental surveillance and control. Each would require some restructuring through mergers and legal arrangements to provide the necessary concentration of activity and governmental authority.

Among these three alternatives, an intergovernmental arrangement seems dictated by the existence of state enterprises in the petrochemical industry; but a Latin American multinational enterprise could be formed out of the local suppliers in the automobile industry. However, in either, the projected need for joint ventures with foreign companies raises some legal as well as operational problems.

The use of the international companies as the means of coordination and implementation would provide some substantial gains in economizing on management, in assuring a continuing flow of technology and skills, and a potential expansion of exports. If, for example, either industry were restructured to create, say, three international companies operating over the region, some competition could be retained among these companies while at the same time there would be sufficient economies of scale to achieve worldwide competitiveness in some of the major products. Exports to third markets could be expanded, and the arrangement could help direct the benefits of these sales to particular members if the criterion of balance of payments required it. These companies also would be able to respond readily to requirements as to how the technology ought to be balanced among members; they could shift products and thereby alter employment as required, and they could help train local managers and workers in the requisite skills.

Use of the international companies as the vehicle of implementation would require some specific arrangements with the supplier companies, and joint ventures would probably be the appropriate answer. This would mean joint ventures of the international companies with supplier companies (some locally and some foreign owned) in automobiles, and with state enterprises in petrochemicals. In both industries, such joint ventures already exist in a few countries; the experience would not be new. There are some trade-offs required in the formation of such joint ventures, for the expansion of exports from a joint venture is less attractive to an international company than from a wholly owned subsidiary; but the advantages of control by nationals would probably outweigh the possibility of a smaller volume of exports.

Whether the control mechanism is exercised through Latin American or through international companies, it would probably be desirable to have joint ventures among national and foreign-owned enterprises—if for no other reason than to remove the mystery about what the others are doing under the agreement. To implement such an accord, it would be necessary for governments to oversee intercompany pricing, intercompany charges, the pattern of sales, profit allocation, research and development activities, supply of components, purchase of raw materials, employment policies, and so forth. Continued intergovernmental surveillance over these facets of intercompany operation would probably lead to much sounder policies as to content requirements (regionally based), selection of product lines, and location of new facilities or expansion of existing ones. In addition, participation by member governments would also provide their officials with a better understanding of the impacts of the tax structure and other national governmental controls.

Surveillance does not necessarily imply control, but new governmental controls would be needed. Not all of the above relationships would be controlled; in fact, surveillance and understanding would soon show which aspects require control and which could and should be left to company decision and the market forces. Continued surveillance would be necessary given the anticipated heavy reliance on the international companies and the need of governments to keep informed about the distribution of benefits. But some controls undoubtedly would be necessary. The very fact of a complementation agreement creates degrees of monopoly: under most such agreements, only a few companies would remain to compete in oligopolistic markets. Prices would probably be brought under control by the participating governments; as substantial purchasers of both automobiles and petrochemicals, the governments have both the incentive and the power to control prices. In addition, the necessity of maintaining a balance of benefits through cost reduction would probably lead to some continuing control over prices.

Complementation agreements in any one of various sectors, such as automobiles, and petrochemicals will likely give rise to difficult problems of balancing equity. Some smaller countries, such as Ecuador, Paraguay,

Uruguay, Bolivia, Peru, and Chile, are not in a position to make a significant economic contribution in all major sectors—at best only in a few. If the number of sectoral agreements is expanded significantly, it is highly unlikely that many countries will have the economic capacity to commit sufficient resources to achieve economies of scale in all of them. Specialization among the sectors will be required. The choice, however, need not be on the basis of static, least-cost or comparative-advantage criteria. Rather, within fairly wide limits, the choice *can* be made according to the "balancing" criteria mentioned earlier, without loss of significant efficiency. For example, Ecuador could gain efficiency by accepting a larger benefit under one sectoral arrangement than in another; it might trade off a potential balance-of-payments benefit under one for a higher technology under another. These considerations would lead to a balancing of benefits among various complementation agreements, rather than within each agreement. Balancing the benefits for each country within each agreement would make it more difficult—or even impossible in some instances—to gain the desired economies.

Since these trade-offs among agreements would have to be made within a period of time that made each member feel secure in the equity arrangements, there would be need for a Latin American organization to make the various arrangements and facilitate the required trade-offs. Each member nation would undoubtedly want to participate in each of the key industries, so as not to be left in a technological backwater.

As a consequence of these decisions and trade-offs, a pattern of specialization would develop, though not necessarily that which would have arisen under free trade. The extent of the divergence from free trade would be determined by the degree to which the weight of "historical accident" and an "early start" would be modified by negotiations, taking into account the dynamic factors that shape the growth of industries. The objective of the negotiations would be to achieve a degree and pattern of specialization that would permit worldwide competitiveness while satisfying national interests, including the desire for reciprocity and equity.

Though the objective is clear and the mechanisms are available, the road is still not readily travelled. There are at least four obstacles to industrial complementation agreements created by various physical, economic, political, and other circumstances. One such obstacle is the inadequacy of transportation; another is instability in exchange rates; a third is intervention by foreign governments (mainly the U.S. government); and the last is the credibility and commitment of Latin American governments.

An efficient transport system is required. The trading of components and materials and final products among interlocked entities requires not only a smoothly functioning transport system but also a low-cost one, to keep the final price competitive worldwide. Evidence gathered on the automotive and petrochemical industries is that transport facilities can be developed for intraregional trade, but they are costly now—more so than in international trade for the normal routes move out from the region. Regional costs need to be lowered; this can be done not only by improving facilities but also by breaking the administrative logjams and bureaucratic procedures which require multiple copies of documents and long delays in clearances through customs. The Latin American countries are working on both a more efficient transport system and faster clearances, but much needs to be done to assure that the specialization under complementation agreements is not ruined by transport costs and delays. Of course, some pressure would be built up by the agreements themselves to hasten reforms in the transport system; as soon as governmental agencies saw what was being done under the integration schemes, they might be induced to move more rapidly in the transport field.

The effects of exchange rate instability are more complex and difficult to resolve. The causes of instability include both political and macroeconomic forces. But instability is unacceptable if industrial integration is to occur under complementation agreements. This fact has been recognized readily under the European Community (EC) agricultural accord and is seen from the fact that price arrangements will be altered by changes in exchange rates. But, more importantly, the sharing arrangements would be shifted with each change in exchange rates. If one country were to change its exchange rate, the prices of the components it contributed would change; and if it were producing a component also manufactured in another country, a shift in sources could arise. To the extent that such a shift is prohibited by some rule, the devaluation of the currency would be ineffective; to the extent that the devaluation were permitted to take effect, sharing would be disrupted.

If exchange rates are not to be held stable, complex regulations will be required to make certain that the shares in *real* terms are not altered by shifts in these rates. To talk of equitable sharing and the allocation of production and prices, costs, and so forth, and not to take into account the condition of the exchange market, is to court continued conflicts among members and an eventual fracturing of the agreement. For Latin America, this dilemma seems an almost impossible hurdle.

The third obstacle arises from the fact that the U.S. government has continued to intervene in the affairs of affiliates of the international companies headquartered in the United States. The three areas in which governmental interference occurs—export control, antitrust, and balance of payments—are all vital concerns of Latin America.* Suppose, for ex-

---

* For analysis of these interferences, see Jack N. Behrman, *National Interests and the Multinational Enterprise* (Englewood Cliffs: Prentice-Hall, 1970), Part II.

ample, that the members of an integrated industry wished to export to Eastern Europe or other areas that are under U.S. export controls; they might be prohibited from doing so if the U.S. government decided to extend its authority through the U.S. parent company. The complexity of the controls is greater and their severity more intense if the transactions are with Communist China—a country with which many Latin American governments still do business. Any complementation agreement including U.S. companies would provide a much larger area for the extension of American controls than existed previously.

Or suppose that the agreement required the merging of several U.S. companies with European subsidiaries in Latin America or with local companies, and the result was a monopolistic arrangement. The U.S. Department of Justice might consider the arrangement illegal and prosecute not only the U.S. parents but the European companies (through their U.S. affiliates) for injuring U.S. commerce. The agreements would seem to require the prior approval of the U.S. government; procedures for such prior approval have not yet been established.

Again, suppose that the complementation-agreement authorities determined that larger sums should be reinvested for future expansion—at a time when the U.S. government was insisting on a larger return of earnings from overseas affiliates to ease its payments deficit. The difficulty of resolving this conflict would be increased by the fact that the U.S. affiliates would be locked into a large integration scheme with substantial Latin American interests.

A final obstacle exists in the feeling among Latin American countries that governmental commitments are not necessarily credible. Too often government officials have violated their oral or written agreements or withdrawn a commitment. The large investments needed to implement complementation agreements will be difficult to induce in an atmosphere of governmental reversals.

Despite the complex problems of balancing equity and efficiency and of making trade-offs between national, regional, and international interests, there are means of creating successful groupings in major industries—as shown in the succeeding analysis of automobiles and petrochemicals. The minimum prerequisites are that at least the three big countries accede to the agreement, that there is an agreement on the location of production and patterns of trade within the region, that the problem of exchange rate instability is met, that means are found to circumvent potential interference by the U.S. government, and that the governments entering into the pact stick by their word.

Given these conditions, the mechanisms are available, and all that awaits is the initiative on the part of governments to provide to industry the guidelines as to what an agreement must contain. Then it could be left to the industry to take the initiative in making proposals. To date the participation of industry has been more that of making certain its own (company or national) interests were protected, rather than seeking ways of creating integrated industries. This is largely because there are no adequate guidelines from governments on the tough issues of sharing, allocation of industry, and the control mechanisms. Were such guidelines to be established, progress could be made.

However, the attitude of the major governments—Mexico, Brazil, and Argentina—is that they can go it alone or that each would be better off to "integrate" with the U.S. economy rather than with their poorer neighbors. It is highly unlikely that they will make the sacrifices of time and effort to produce the necessary guidelines. And company officials can be expected even less to make the effort. We can expect continued verbal assurances of dedication to integration, but few concrete steps are likely to be made involving sacrifices by the larger countries to the smaller or sacrifices of national objectives to regional interests. Either the major countries will make the leap to industrial efficiency in specialized fields on their own, or Latin America will remain in the industrial backwater, lagging further behind the United States and the other more-advanced countries. In the latter event, it can be expected that tensions between the two groups of countries will continue to rise from the fact of this lag.

The same conclusion appears to be sustainable for the Andean subregion. For the subregion to be viable, Venezuela needs to be in; it is not. And conflicts of interests between the relatively outward-looking economies of Colombia and Ecuador and the relatively inward-looking ones of Peru and Chile appear likely to become more intense, fracturing their recent accords. To inject a more hopeful note in this pessimistic assessment, it appears that an initiative by the advanced countries might provide the catalyst to induce use of complementation agreements to achieve both efficiency and equity. The initiative would be in the form of an invitation to negotiate intergovernmental agreements covering specific industrial sectors among the advanced and developing countries. The agreements would utilize the existence and capabilities of the international companies, without which several industries in the developing countries probably cannot become competitive internationally. These arrangements would remove the pressure for generalized preferences and make certain, through an agreed-upon pattern of production and trade, that the preferences extended under the agreements were effective.

Such an initiative on the part of the developed countries does not seem wholly outside the realm of the possible; it would follow the European Coal and Steel Community, the pattern of the United States-Canadian auto agreement, and the suggestion by Eric Wyndham White that further progress in industrial integration is likely to come through sectoral arrangements. Europe is itself discussing just such a move under the Colonna Plan, and the United States is inching toward such an arrangement in its continued pressures for regularization of trade in textiles. The larger the number of countries and the number of sectors covered, the more readily would the twin objectives of efficiency and equity be achieved.

*   *   *   *

In the course of a field trip to the region during June-July 1971, after the above was written, certain trends of interest were observed, some more encouraging than others for the development of regional integration. The use of complementation agreements was found to be increasing, with six-teen agreements in various stages of negotiation beyond the nine agreements already in operation, and the seven that have been signed but are not yet functioning. Within the Andean group, the commission that was established under the Treaty of Cartegena has recognized the importance of pressing quickly for an industrial policy that puts a high priority on sharing indus-trial activities among the member countries so as to assure "equitable" participation in industrial growth. Some twenty industrial sectors, including autos and chemicals, were held out from the tariff-reduction procedures to permit special treatment. However, no procedures have as yet been adopted for the allocation of specific activities among members.

In the auto industry, some of the international companies are proceeding to integrate still further their affiliates' operations throughout the region by arranging international swaps under national provisions permitting imports of components to the extent of exports. So far, the governments have been unwilling to agree on multilateral rules for such swaps or to accept the concept of "regional content." To the contrary, the expansion of the auto industry on a national basis is being promoted by a number of governments, including Brazil, Mexico, Argentina, Colombia, Venezuela, Peru, and Chile. Brazil considers that its internal market is sufficiently large that the industry can produce cheap transportation for domestic consumption and eventually export. Mexico, in an effort to increase production volume and thus achieve lower unit costs, has raised still higher the percentage of components that auto companies must export. The other countries are seeking to rationalize domestic production by controlling or reducing the number of companies in operation, and some are restricting the number of models to be produced.

The future position of international companies, in general, including their capacity to promote regional integration, is being significantly affected by the continued trend toward forcing the "localization" of ownership of such companies (except in Brazil, at the moment). In Colombia, Peru, and Chile, requirements have been imposed or are anticipated requiring foreign auto companies to enter into partnership with local companies or with the government. Mexico has had an on-again-off-again policy of encouraging or requiring a minority (or majority) local ownership of industrial facilities. Argentina is tightening its policy in the same direction. Venezuela in 1971 passed a law calling for the reversion of all oil concessions to the state in 1980, including all facilities built on the concession lands. And the Andean group has recently adopted an investment code that requires mixed ownership of all foreign-owned affiliates within a period of about fifteen years.

The Problem: Inadequacies of Tariff Reductions

The major technique of economic integration has been that of tariff reductions. The unification of Germany began as a customs union; the thirteen American colonies rejected duties among themselves to foster integration; both the European Common Market and the European Free Trade Association (EFTA) have used tariff reductions as the first step toward economic integration, and LAFTA has followed suit. The Montevideo Treaty provided three means of removing tariffs among the members, and recently a fourth tactic has been adopted; these are described below. But each has been found wanting as a means of achieving agreement among members to permit the free movement of goods among themselves. The reasons for their rejection point the direction to a new approach.

The Latin American governments recognized that their situation was different from that of Europe in that members would not accept a fixed timetable and formula for elimination of duties. There was too great a reluctance to fix formulae, since they might not provide for reciprocal benefits and concessions. To provide for some automaticity in tariff reductions, so-called national lists were established under which each LAFTA member would reduce by 8 per cent each year for twelve years the weighted average of duties collected on imports from all other member countries. Each member could select the items on which the reductions were made, and if it wished to retract the duties reduced it could do so, substituting equivalent benefits. With continued progress each year duties on all trade existing in 1960 would be removed by the end of twelve years.

To make certain that all other traded items (i.e., those that had not been the subject of trade before 1960) were without duty by 1973, a common list was established under which members would negotiate the items which would be given free-trade treatment. The negotiations were to proceed in four stages, each to cover 25 per cent of all items traded intraregionally. By 1973 "substantially all" traded items were to be accorded free entry, but none of the reductions proposed would be put into effect until that date. The only effective reductions prior to 1973 would be those under the national lists, and reductions on any item were revocable.

To obtain additional immediate duty reductions and include items that were not necessarily currently traded, but which had a trade potential, provision was made for duty reductions by industrial sectors. These sectoral arrangements are called complementation agreements. The treaty provided that any two countries could enter such an agreement on any number of products within an industry to reduce duties on these items. The purpose was not only to accelerate reduction of duties but also to induce investment to serve the wider market created. The reductions under these agreements would be placed on the national lists of signatory countries, opening their markets freely to all LAFTA members. Later, the conditional most-favored-nation form had to be substituted, restricting the benefits to signatory countries.

Because member governments have been reluctant to open certain sectors to free trade, there has arisen also a form of temporary complementation agreement, the *acuerdos de excedentes y faltantes* or agreements on surpluses and deficits, referred to earlier. They are a means of providing for the reduction of duties on items in surplus in one country and in deficit in another, thus sopping up overcapacity. Once the deficit is removed in the importing country, or surpluses no longer exist in the exporting country, the agreement is terminated.

National Lists

The objective of the national lists was to provide immediate reduction of duties on a most-favored-nation basis so that the widening of markets would induce establishment of new plants and expansion of existing facilities on a more economic scale. But to obtain this result would have required some assurance that duties would be reduced to levels permitting trade within a short (investment) period. There was no assurance that an item receiving a reduction this year would be further reduced in any subsequent year. Nor was there assurance that the reduction would not be revoked when the granting country obtained additional production capacity within its national territory.

Given the substantial investments required in automobiles and chemicals, for example, this lack of certainty prevented any significant "integrative" response. Chemical industry officials in Latin America have remained reluctant to risk a $10–$15 million investment on the basis of a commitment that the governments could revoke on short notice. For example, since the Brazilian Petróleo Brasileiro SA (PETROBRAS) was the only producer of SBR rubber in 1963, it obtained duty concessions from five other members of LAFTA; during 1964 and 1965, SBR rubber constituted the largest single item in the intra-LAFTA chemical trade. But by 1967, both Argentina and Mexico had plants on stream, and both suspended their concessions to Brazil.

This pattern has been repeated on many occasions. The original concession opens a market in the importing country and makes it clear that production is possible (even if somewhat uneconomic). Armed with the information on the market, local interests persuade the government to revoke the concession when a plant is built locally. The national industrial orientation of the governments supports this approach; the advantages of duty reductions are short-lived.

The fact that the national lists are restricted to items already traded has also prevented their being very useful in either automobiles or chemicals, for neither industry showed significant intraregional trade. In automobiles, the local-content requirements have prevented significant imports; what was imported came from outside the region. The introduction of investment in new chemical products has been stymied by lack of a wider market, and the wider market is not likely to be created through the national lists until items are traded extensively. Pressure is needed to put the items on the national lists with one country *asking* for a concession from others; but that country is not likely to ask for concessions when it is not certain that it can produce the item efficiently enough to capture the market.

The national lists also have a weakness in that the more significant or key industries are likely to be the last to receive concessions. Each country will grant concessions that are relatively easy and undisturbing and in which it does not expect to have a production interest. The key industries are those in which the country *hopes* to have such an interest. Few concessions have been granted on chemicals and automobiles. In a sample of 100 chemical products, on which about 1,000 concessions could have been granted, only 17 per cent of the potential concessions had been granted by 1967. Although some 70 of the 100 chemicals were the subject of at least one concession, concessions on 18 were made by one country, 17 by two countries, 18 by three countries, and 10 by four, and 7 by five or more countries. Less than one-third of the total concessions were made by the three largest countries (Argentina, Brazil, and Mexico). Only about 6 per cent of the potential concessions could be said to have created a substantial regional market. Most of the rest of the concessions were by the smaller countries, opening them to the larger countries but leaving the regional market still substantially less than all LAFTA.

Even for those chemicals in which a large market was covered by concessions, the duty reductions by 1967 were not significant. For example, in cellulose acetate, five countries granted national list concessions in 1965, but trade arose only between Argentina and Brazil. The concessions granted by Ecuador and Uruguay left the rates at 30 per cent and 38 per cent respectively; a duty of 30 per cent is considered adequate to preclude imports in this item. This illustrates the absence of a mechanism under the national lists to make certain that the duty reductions were really effective in creating a market of regional size. Cuts in duties were made in proportion to existing national duty levels—not toward some realistic level inducing trade. Even if a realistic level were to be reached eventually, as duties were reduced, there was no certainty that "eventually" is the same year for all—save in the event of the final year. Any country could hold off on a given item until the twelfth year.

Common List

The common list remedied one of the major deficiencies of the national lists in that duty reductions achieved through it were irrevocable. However, the list was to become applicable only at the end of the twelve-year period. Even with this delay, the lesser-developed countries in LAFTA were concerned that the procedure did not guarantee reciprocity to them, for concessions were not negotiated. The procedure simply provided for free trade in "substantially all" trade in 1973 without concern for its impacts on the poorer countries. At the first meeting to select items for the common list, the lesser-developed members insisted on assurances of equitable treatment, i.e., reciprocity of benefits. Other members had to agree that the distribution of benefits would not be left strictly to the market system. It was agreed, instead, that the location of industry under free trade would have to be determined by joint planning so as to achieve an equitable distribution of the benefits of integration.*

This conflict over reciprocity became so intense that the second round under the common-list procedure, to raise the volume of trade encompassed by it to 50 per cent, was not successful. Countries refused to place items on the list without a resolution of the problem of reciprocity— meaning a mechanism to determine the location of industry under integration. At the second round in 1967, LAFTA members decided that the common-list products would not be subject to free trade in 1973, as projected, but only after new negotiation procedures were worked out to provide reciprocity; no date was set for making this determination. This breakdown has caused some observers, such as Raúl Prebisch, to assert that the procedure should be left aside as an interesting experiment that helped define the problems more precisely. There seems to be general recognition that the technique is dead, at least for any relevant future period.

Complementation Agreements

The Montevideo Treaty was amended and expanded in 1964 by a resolution spelling out the objectives of complementation agreements, which

---

* (Resolution 100-IV, paragraphs I.B. 7 & 9).

were intended to supplement other techniques and correct some of their inadequacies. The objectives included extending the agreements to products not yet traded intraregionally, covering all available facilities of production in the sector concerned, and insuring equitable conditions of competition which would also improve quality of products and reduce prices.

The treaty requires that all agreements spell out the products covered and the means, procedures, and schedule for duty reductions. The agreements may merely include tariff reductions or extend to the elimination of all duties and allocation of production among members. The agreements must be open to all LAFTA members, and concessions must be automatically extended to the lesser-developed countries (Bolivia, Ecuador, and Paraguay). The agreements are also supposed to include means of harmonizing the members' treatment of imports from third countries (both inside and outside LAFTA) of similar products or materials and components; treatment of capital inflows and related services is likewise to be harmonized.

Complementation arrangements may be distinguished according to whether they apply to trade among the members for *all* companies in the sector or only among the affiliates of one company across national boundaries; whether they cover *all* items produced in the sector or only selected end products; whether they provide for elimination of duties or only for preferential reductions; and whether or not they provide for production allocation. One agreement may cover all companies in an industrial sector, though they may apply only to a small list of products and provide for reduction but not elimination of duties. Or the agreement may relate to a single company, providing for elimination of all duties but assuring allocation of production among the member countries. Through 1970, sixteen such agreements had been signed.

The *first agreement* was for the benefit of a single company—International Business Machines Corporation—after it promised to establish plants in Argentina and Brazil. Chile joined in order to induce Crown Zellerbach Corporation to set up facilities there; Uruguay also signed. The agreement covered electronic-data processing machines, parts for them, and paper tape and cards for use in these machines. It provided for the removal of all duties among the signatory countries. The result was a volume of trade among these countries in 1967 totalling $4 million, as compared to none before.

The *second agreement,* signed in 1964 by Argentina, Brazil, Chile, and Mexico, provided for free trade in certain types of electronic tubes. Since producers already existed in each country—all subsidiaries of foreign corporations—some exemptions from free trade were granted to permit gradual adjustment to competition. General Electric Company in Uruguay was granted exemptions on fifty-two types of tubes; the plants of the Radio Corporation of America in Chile and Mexico were granted twenty and forty-two exemptions respectively; and the subsidiaries of RCA and N.V. Philips' Gloeilampenfabrieken in Argentina were granted a total of fourteen exemptions. These exemptions were gradually removed so that by 1971 they no longer existed. As a result of the agreement, trade in the region rose from $730,000 in 1964 to $1.6 million in 1966.*

A further stimulus to negotiation of agreements came from the shift from the most-favored-nation clause to the *conditional* most-favored-nation procedure in 1964, which voided the necessity to extend the duty reductions to nonsignatory countries within LAFTA. Thereafter, all but the lesser-developed countries had to buy their way into the agreement with similar concessions. In 1965, some thirty agreements were recommended by private enterprises in as many sectoral meetings, but only two were approved and adopted by governments.

The *third agreement* was a bilateral arrangement between Brazil and Uruguay covering domestic appliances.

The *fourth agreement,* likewise signed by Uruguay and Brazil, related to electrical communication equipment.

By 1967, governments themselves saw the potentialities of substituting the agreements for the stalemate on the national and common lists. Government-initiated negotiations resulted in two agreements affecting chemicals and petrochemicals.

The *fifth agreement* (1967) was ratified by all LAFTA members except Venezuela. The eight relatively more-developed countries agreed to extend irrevocable duty reductions on a wide range of chemical products, and the reductions were extended freely to the three lesser-developed countries (Bolivia, Ecuador, and Paraguay). However, the agreement included only nine concessions on petrochemicals—compared to at least a hundred that, according to LAFTA estimates, require regional markets. In addition, six of the concessions had already been granted through their having been placed on the national lists of the countries involved; therefore, only three were additional. Further, there were no provisions for harmonizing treatment of imports from third countries, the inflow of capital investment, or imports of services related to these products. The agreement was simply a tariff-reduction agreement on a very limited number of petrochemicals.**

---

* Data are difficult to obtain on Latin American trade. Though extensive data exist, they are often contradictory and are seldom very recent. Most of the data for these studies relate to the years of the mid-1960's, which were the latest available in 1969–70.

** Reductions were granted also on a variety of chemicals not in the petrochemical category (i.e., inorganic chemicals and vegetable oils) that do not require large-scale production and can be produced efficiently for small markets.

It did nothing to assure integration of the industry, and provided for no incentives to restructure the facilities in any direction.

The *sixth agreement* came under the Andean Pact, signed by Ecuador, Peru, Chile, Colombia, and Bolivia. It provided for free trade and a common external tariff on a large number of petrochemicals. The Andean agreement also provided a complete plan for the development of the various branches of the petrochemical industry, with planned investment, coordination of production policies, and agreement of location of plants among members. This agreement was much more in line with the intent of the 1964 resolution on complementation agreements. It provided for allocation of production among the members, thus eliciting a given integration structure in the industry. It also provided for elimination of duties completely by 1973 and for common external tariffs in a range of 15 per cent to 50 per cent on the products covered. Further, the agreement extended concessions on petrochemicals that need an intraregional market to permit efficient production—items such as styrene, hexamethylene diamene, and vinyl acetate.

The *seventh agreement,* signed in 1968 by Argentina and Uruguay, covered certain home appliances and overlapped to some extent the items in the prior Brazilian-Uruguay agreement on electrical appliances. Thus, some of the items are traded freely among the three countries, and others only bilaterally between two partners.

The *eighth agreement* (1969) covered a wide range of glass products traded between Mexico and Argentina. (Brazil, Chile, Colombia, Peru and Uruguay participated in the negotiations but did not sign the agreement.)

The *ninth agreement* was negotiated by Brazil and Mexico covering equipment for the generation, transmission, and distribution of electricity.

The *tenth agreement,* signed in June 1970 by Argentina, Brazil, and Mexico, covered a substantial number of office machines. The agreement provided for establishing margins of preference compared to other countries, and these margins were supposed to be maintained; a further undertaking was to seek harmonization of barriers to third countries so that preferences are similar. The agreement could be renounced by any party after two years, with concessions to remain in effect for one year after denunciation.

The *eleventh agreement,* signed by the same three countries with Chile, took into account Chile's special situation in this same sector. It permitted a more gradual tariff-reduction procedure and covered only 4 items and 21 concessions, compared to the 32 items and 51 concessions under the tenth agreement.

The *twelfth through the sixteenth agreements* were signed in the last months of 1970. The additional five agreements covered the sectors of petrochemicals; phonographic equipment; pharmaceuticals; electronic and electrical communications; and refrigeration, air-conditioning equipment, and electrical domestic appliances. Brazil and Mexico are in all five; Argentina is in three, Venezuela two, and Uruguay one. The phonographic-equipment agreement covered only five countries and 5 items for 27 concessions; electronic and electrical communications two countries and 41 items for 83 concessions; refrigeration-air conditioning-domestic appliances two countries and 25 items for 42 concessions. Petrochemicals involved 300 items among four countries, but only 67 initial concessions, with others to be negotiated later. The pharmaceuticals agreement involved 300 items among three countries for 439 concessions.

A large variety of other agreements have been proposed, both by private industry and governmental bodies, but the initiative has been largely taken by industry associations. They cover products in the industrial sectors of paper and pulp, machine tools, textile machinery, pulp and paper machinery and equipment, bakery and confectionary machinery, electric lamps and illuminating units, medical instruments and equipment, photographic equipment, cosmetics and toiletries, and canned fruits and vegetables. Meetings on these agreements are sponsored by LAFTA, and some of the sectors have held six or more sessions, while others are in the first stages of negotiation. In the main, the negotiations cover only a limited number of products, though some run to more than a hundred items in a given sector. Some other negotiations have run into impassable obstacles and have been disbanded.

Among the above there are none either signed or in negotiation concerning the automotive industry. Conversations have been conducted among various companies and associations, and proposals have been made on a tentative basis to particular governments, sounding out their interests, but nothing has been made public on these efforts.

The most recent agreement, signed by Argentina, Brazil, Mexico, and Venezuela in the petrochemical sector, could provide the basis for a substantial integration among the larger countries, but the members are reluctant to include significant items as yet. The Andean (sixth) agreement is the only one that attempts a solution to the inadequacies of mere tariff reduction that have surfaced in the development of LAFTA. But it faces many problems in being implemented successfully, as will be analyzed in subsequent chapters.

Inadequacies

The major inadequacy of tariff reduction or free trade as a means of economic integration is the absence of a procedure for balancing the benefits of integration among the country members. This problem can be overlooked by countries that see their own growth moving as fast as the country can sustain, whether prior to or during the integration process. Such was largely the case in Europe during the first decade of the Common Market, from 1958 to 1968. In such a setting, it is relatively easy for the negotiating countries to put the problems of integration on the other side of the table, as it is said, and face them together. The division of the benefits then is less of a problem, for all anticipate a rapid growth rate. A Social Fund was provided under the Treaty of Rome to ease some adjustments that were anticipated, as was an Investment Bank to help some lesser-developed regions. Another fund for adjustment to trade shifts was provided, but it was not drawn upon. The continued growth in Europe permitted ready adjustment to pressures of the opening of trade.

When integrating countries are not in so favorable a situation, there is considerably more concern over the division of benefits; this has been true in Europe regarding policies toward agriculture, a sector with much less growth potential. The more-advanced countries see their growth being slowed by the necessity to bring the slower countries along, and the latter see their resources and future potential being pulled into the economies of more rapid growing countries. The problems of integration, therefore, are placed in the middle of the table, with the parties confronting each other over the division of benefits.

The nature of the difficulties in the way of economic integration in Latin America can best be understood through a brief review of (1) attitudes toward the common and national lists, (2) the bias toward national orientation of industrial policy, and (3) the fears of what might happen under free trade.

The attitude toward the common and national lists was evidenced in the initial rejection by the Latin American countries of a scheduled reduction of all duties, patterned after the Common Market treaty. Rather than agree to a fixed timetable and stepped reductions across the board, the Latin American countries insisted on negotiation at each step in duty reduction. The objective was to make certain that the benefits balanced, for it was conceivable that the excessively high duties on some items would not be reduced to a point where trade could take place until the last years of the transition.

The demise of the list approach was caused by the difficulty of finding a balance among the concessions offered. It was easy to move through the first stages for the concessions granted were not considered by most countries offering them to be significant. But, as the lists expanded and items had to be included that might alter trade and production patterns, the burdens became apparent and the benefits were not at all clear. Under the national lists, the reaction to market pressure from a duty reduction was frequently to remove the concession. The common list simply failed to be negotiated for the second 25 per cent round.

The lists were abandoned also because the complexity of trying to determine the balance among the benefits of a variety of concessions offered by each country proved overwhelming. The more limited complementation agreements provided an escape from this complexity by permitting negotiation on only a few items by only a few countries; it was easier to balance benefits.

A second evidence of the inadequacy of the list approach—and the reason for the slow negotiation of the complementation agreements—has been the continued *national* orientation of industrial policy. Almost every country in Latin America feels the need to accomplish certain national integration goals before opening the economy to regional integration. There is still a need to shift labor from agriculture to industry, to improve productivity in both agriculture and industry, and to broaden the industrial base. Each country feels that it can do this more successfully in a relatively closed economy, and each fears the interference in these decisions that would arise under regional integration. It is only from a strong domestic situation, buttressed with many industries and competitive operations, that each feels it could be certain of gaining an "equitable" share of the benefits of integration.

This desire to bargain from strength is heightened by still another inadequacy in the tariff-reduction approach to integration, namely, the prospects of what might happen under free trade. Rather than the perfect adjustment often claimed for free international markets, experience indicates the balance of benefits of production and exports would be substantially skewed in favor of the already advanced countries. Specific evidence is found in both the automotive and petrochemical industries.

Although there is neither a complementation agreement in automobiles nor a substantial reduction of duties, there is an arrangement between Chile and Argentina under which each adopted national legislation permitting some parts imported from the other to be considered as national content (and thereby free of duty), provided that an equal total value of parts was exported to the other. The annual volume of such swaps was limited to 6 per cent of domestic content in Argentina (the larger market) and 20 per cent in Chile. In addition, only 30 per cent of a company's requirements for an individual part could be imported from the other country; however, this restriction was removed completely if the part had been previously entirely imported under the former local-content requirements *or* if affiliates of the same company shifted production to gain economies of scale, producing all of one item in Argentina and another in Chile.

Although this arrangement between Chile and Argentina is less than free trade by a large margin, the results point toward what would happen under free trade. Bilateral exchanges rose from zero in 1965 to $12 million in 1968. Argentina's exports to Chile consisted mainly of engines and transmission components. Chile exported a variety of smaller products—radiators, steering wheels, cables, heaters, wheels, and springs. The exchange was initially heavily in favor of Argentina because of the more-advanced nature of its industry. But with governmental incentives, Chilean industry expanded production fairly promptly.

A second problem faced Chile because of the discrepancies in the scale of production in the two countries. Chilean production of vehicles had been on the order of 20,000 per year, compared with 200,000 in Argentina. An Argentine company producing a part for Chile could expand operations by only 10 per cent, while a Chilean firm was required to expand tenfold to serve the Argentine market. When the 30 per cent limitation applied, the expansion was only 3 per cent for the Argentine company, but it was still threefold for the Chilean, requiring a substantial increase in investment. Uncertainty about the continuation of the arrangement caused Chilean companies to delay making investments.

The arrangement had serious impacts on prices of components. For example, one auto company shifted the production of springs from its Argentine affiliate into its Chilean affiliate in order to generate Chilean exports and expand Argentine exports of motors. As a consequence, the cost of springs to the Argentine assembler rose 90 per cent above the previous level, increasing the total cost by $750,000 annually. This increase was offset by the reduced costs of production of the motor resulting from the 10 per cent increase in volume due to Chile. Without this arrangement, both items would have been produced in Argentina. Another firm reported that the average cost of parts obtained from Chile was three times the Argentine cost for the same items. Again, under free trade the shift would not have occurred. For a country that is insistent on obtaining a foothold in the automobile industry, the free trade result is unacceptable. Tariff reduction will be accepted *only* if it permits industrial development consistent with the national interest as seen by national governments.

Peru has embarked also on a program of expanding the auto industry, raising domestic content. Estimates by company officials are that a Peruvian regulation requiring 50 per cent of components from *Andean* countries would raise cost by between 100 and 200 per cent of U.S. costs; if the requirement were for Peruvian parts, the costs would be even higher. Even within an Andean agreement, the officials estimated that Peru would not be competitive internationally or regionally if Peruvian suppliers were required. The only way in which Peru might be competitive would be if it could serve the entire Latin American market in a few restricted parts.

Similarly, economies of scale and existing overcapacity situations lead company officials to estimate that one Colombian producer of castings would supply all the LAFTA countries with engine blocks and still not use all his existing capacity; the company was then operating at only 30 to 40 per cent of capacity (selling other castings mostly in the national market), but the government showed no interest in a complementation agreement. Another auto supplier was seeking to produce generators and electrical equipment on a scale that required serving the LAFTA market and exporting to Africa. A Venezuelan manufacturer of frames was already exporting to Chile and was planning capacity to serve both Mexico and the United States.

These examples point to the conclusion that existing capacities would require an opening of the entire LAFTA market to reduce costs of production for a few companies *and* that free trade would settle production where it already exists, without reference to the interests of each country in its participation in the industry. To offset the potential results under free trade, each is building up its own facilities, which in turn are becoming obstacles to future agreement on the reduction of barriers and are creating costly overcapacity. Obtaining a strong national bargaining position may become so costly that no government will feel capable of abandoning a past investment to adjust to the integrated pattern.*

There is an urgency, therefore, in meeting the problems of integration before the hostages created by national industrial policies become too numerous and costly to buy back.

Similarly, in the petrochemical industry, the move to free trade would produce a structure of Latin American industry wholly unacceptable to the national governments. Although analysis of the costs and facilities in Latin America show substantial reductions in costs in an integrated market, the larger cost reductions occur in the medium-sized countries. Colombia was found to be the most efficient producer of a wide range of petrochemicals; in only five major chemicals does it have the lowest potential cost of production at a regional scale of production.**

Among the larger countries, Mexico is the least-cost producer in petrochemicals. Chile is the highest-cost producer, even on a regional scale, with Argentina the highest-cost producer among the larger countries. The reason for the Argentine position lies in relative high costs of electric power, labor, sugar, phosphate rock, and calcite. Brazil has the highest petrochemical raw materials prices in Latin America, with the exception of Chile; Mexico has higher labor and electric power costs than Brazil, Colombia, or Peru; and its continued high labor costs prevent it from displacing Colombia as the least-cost producer. Venezuela has the lowest-cost raw materials for petrochemicals, but its other inputs are high cost; its

---

* See Appendix A.
** See Appendix B.

labor is seven times that of Peru. In sum, no country has a corner on the lowest-cost inputs, but Colombia is in a more favorable position than any other.

When the analysis is broken down by commodity groups, the positions shift somewhat according to whether labor or natural resources are more important in production and the relative costs of each in the country. For example, Venezuela has the best endowment of raw materials; despite the highest labor cost, it has a less-than-average production cost for organic chemicals. Even Chile, which has a generally unfavorable cost position, is in a strong competitive position in basic acids because of the abundance of nitric and cupric raw materials. Peru holds a similar position to Chile. Although Argentina is not competitive in most chemicals relative to the rest of the region, it is relatively better in the production of olefins and industrial alcohols as a result of the relatively low price of its natural gas.

Under current conditions, if integration occurred in the industry, production would tend to shift to Mexico and Colombia and out of Chile, Peru, and Argentina, with Brazil and Venezuela taking intermediate positions. Ecuador and Bolivia would not acquire any production. The precise commodities in which they would tend to specialize would depend on the aggregate demand. But such a result is wholly unacceptable to Brazil and Argentina and probably even to Venezuela—to say nothing of Peru and Chile, both of which are bending their efforts to create production facilities in the industry. Each wants to share in the use of advanced technology in certain items, rather than produce those using less-advanced techniques.

At the same time, Latin American officials recognize that they must make their industry competitive in the world market, in order to pay for imports needed to accelerate economic growth. They are, therefore, caught in a dilemma in trying to determine the structure of industry under integration. They do not like the potential free-trade solutions; they refuse to accept the determinations made by the international companies; they do not want dictation from international financial organizations as to the multilateral or integration projects they can or should have; yet, the approach decided upon must lead to competitive abilities vis-à-vis the world market. Obviously, new approaches are needed if criteria of balancing of benefits ("reciprocity" or "equity") are to be met satisfactorily and world-market competitiveness is to be gained at the same time.

Arrangements Needed

Any new arrangement must meet at least four conditions. It must provide for an equitable allocation of production among the members (repartition), assuring appropriate volumes of local investment and employment. It must obtain a reduction of costs compared to the prior situation. It must expand the level of trade not only intraregionally but also to extra-regional markets. And it must provide for an equitable sharing of access to advanced technology. In sum, while seeking a more efficient structure of production, an equity solution must also be gained.

In the attempt to achieve these goals, complementation agreements, negotiated along the lines of Resolution 99 adopted in 1964, could have several advantages.

In the first place, if negotiated according to the expectations of their proponents, such agreements permit the repartition of facilities among member countries closely in accord, it is to be hoped, with dynamic or long-term comparative advantage so as to reduce production costs. But this repartition is subject to negotiation and to the determination by experts what the comparative advantages will (or should) be. The trade-offs between an equitable allocation and the more efficient allocation would be negotiated rather than left to the market.

Second, the agreements are expected to avoid overcapacity through the erection of duplicate facilities. Since investment would tend to be allocated according to the decision on repartition, new facilities would be discouraged unless they fit the plan.

Third, there would be a stimulus to attraction of new investment in the industry as a result of the wider market and the certainty of sales from a given facility; this situation reduces the risks of investment and makes it more attractive.

A fourth advantage is that each participant would be able to have a greater diversity of industrial activity than that obtained through serving only the domestic market (on an uneconomic scale) *or* through specialization under world free trade. This, of course, presumes a variety of such agreements, restructuring several industries at the same time.

A fifth advantage lies in the fact that the countries could collectively enter the most advanced industries—such as electronics—and employ the most advanced technologies through serving the larger market and achieving the economies of scale. In this way, they would be able also to compete on the world market.

Finally, the agreements would set up poles of industrial development to which other industrial facilities would be attracted. Each major facility established would require supplier, distributor, and customer services as well as intrastructure. Each would build industrial activity directly and establish more soundly the base for other industry in the area.

In order to achieve these objectives as fully as possible, it was recommended in Resolution 99, though not required, that these integration agreements also include provisions for making the concessions irrevocable, for settlement of disputes, administration of the agreement, harmonization of treatment of imports from third countries of items included in the agreement, incentive programs to implement the agreement, harmonization of treatment of foreign capital inflows, and regulations concerning illegal commercial practices.

3 The Industries: Structures and Alternatives

Following their national orientation to industrial development, Latin American governments have induced investment in both automobiles and petrochemicals without sufficient attention to capacity and competitive ability. Facilities have been duplicated so that regional integration will require substantial shifts in the location of production if efficiency is to reach international levels. These shifts will also require changes in the patterns of ownership, in directions and volumes of trade and financial flows, and in price structures. Though the benefits of regional integration in both industries are substantial and remain enticing, the alternative routes to more efficient production remain confusingly complex and unclear as to their impacts, so that the trade-offs are not readily discernible.

The Automotive Industry

High tariffs and local-content requirements forced the international auto companies to establish facilities in Latin America and help build up local suppliers of parts and components. The result has been overcapacity and high-cost production throughout the region. Several alternative means of rationalization are available to governments, but none seems to be so clearly advantageous as to dictate a policy response.*

The picture of the Latin American automotive industry may be summarized briefly according to location of production, ownership, degree of vertical integration, trade, and prices.

The major centers of production are Mexico, Brazil, and Argentina, together comprising over 90 per cent of the total for Latin America; Brazil produced nearly half of the total of 770,000 units in 1969. Within the major countries, output was by no means distributed evenly among the assembling companies. In Brazil, Volkswagen do Brasil produced nearly half the total output; no other company had annual runs of over 50,000 units, and most were considerably smaller.

Except for three minor companies, all of the assemblers are foreign-owned. The three major U.S. companies—General Motors Corporation, Ford Motor Company, and Chrysler Corporation—are represented in all

---

* A detailed description and analysis of alternatives is provided in Appendix A.

six Latin American countries that produce automobiles, and American Motors Corporation is in two countries. Volkswagenwerk A.G. and FIAT S.p.A. are in four countries, and each of the other European manufacturers is in only one or two countries. Toyota Motor Company and Nissan Motor Company are in three countries each. Among the suppliers of parts and components, many of the North American or European suppliers are represented. Although there are a large number of locally owned suppliers, many of them have minority foreign interests or are licensees of foreign companies.

The pattern of ownership that has developed (or has been required by the host governments) prevents a significant degree of vertical integration in the industry. Local suppliers must provide all but a few of the components of the final vehicles; assembling companies (owned by the major international companies) generally are permitted to produce only engines, major stampings, and transmissions. In most countries, ownership by assemblers of supplier companies is also prohibited. Imports from the parent company are the major means of achieving any vertical integration.

Local-content requirements restrict the volume of imports permissable, and these restrictions are relieved only by the bilateral trade arrangement in auto parts between Argentina and Chile that has been described. As a means of reducing costs, host governments have urged assemblers to export vehicles or parts to other affiliates, to other Latin American markets, or back to parent companies. To induce a greater effort, production quotas or import limits are raised according to the volume of exports achieved.

Because of the restrictions on location of production and the level of output, prices in Latin America remain substantially higher than in the advanced countries. A Volkswagen in Brazil, for example, sells for the equivalent of $2,650, compared to $1,295 in Germany; the compact automobiles produced by GM and Ford also retail for about twice the U.S. levels. Prices of parts and components range from equal that in the United States to as high as five times. The highest prices are found in Chile, where vehicle prices are three to four times world levels and components three to five times. Venezuela has kept the local content down to about 40 per cent so prices are only between 30 to 50 per cent above world prices of vehicles.

Benefits of Integration

The benefits of integration in the auto industry are potentially high since substantial trade is likely to be created, the industries are largely duplicating and competitive, and import duties have been quite high. Given this situation, rationalization of production over the region would produce substantial cost reductions through economies of scale and specialization. To achieve the economies of scale would require, however, fairly large agglomerations of facilities. The production runs for optimum economies are about 500,000 units for engines, ranging upward to over 1 million for stampings. Production runs of 20,000 engines, for example, would raise costs to 50 per cent above those at 500,000. Economies of scale in minor parts can be reached earlier, and the ability to serve several companies simultaneously increases the possibilities. Taking all elements of production into account, Latin American producers have estimated that scale factors alone account for over 50 per cent of the cost differential between U.S. and Latin American operations.

Although costs of raw materials remain high in Latin America, a concentration of production and demand for them would also reduce costs to motor vehicle manufacturers. Capital and total labor costs would not necessarily be affected significantly by sectoral integration.

Another potential benefit, however, is that of introducing more effective competition into the supplier segment of the industry. As it now stands, some countries have only one supplier of a component; its monopoly position tends to raise prices of components and costs of vehicles. With the opening of trade, not only would the supplier segment be rationalized but it would also be faced with competition.

The necessity to reach the economies of scale attainable at production levels of over 200,000 vehicles per year per company means that the number of international companies surviving in the integrated regional market of Latin America would be cut to something under six, compared to the present fourteen. Such a drop also would have benefits in terms of managerial and technical expertise. But the centers of decision would shift in ways not necessarily desired by the host governments.

Alternatives for Governments

In order to achieve competitive efficiency in production of automotive vehicles, Latin American governments have at least four alternative routes: (1) specialization within the international market, producing components or special vehicles for export through various international arrangements and assembling final vehicles locally; (2) specialization within the regional market of Latin America, providing a preferential market for Latin American producers, and exporting some models or components as competitive levels are reached; (3) specialization within a subregional grouping to achieve international competitive levels; and (4) restructuring of the national industry so as to reach least-cost operations within the small national markets by reducing the number of companies involved and the number of models produced.

International Specialization. No country to date has opted for the pattern of international specialization in which it produces only parts of automotive vehicles for the world market. A few produce components for international distribution but also assemble the final units within their own economies—Belgium and South Africa come to mind. But, countries that *can* produce are not satisfied to remain on the fringes of the industry: The automotive industry carries prestige, is a prime pole of development, provides important backward and forward linkages, and provides attractive opportunities for employment. Not even Canada was content to remain a specialized supplier of components to the North American market; it wanted complete production facilities.

Specialization can involve exchange of vehicle types or models, of course, with each country exporting completed or knocked-down vehicles. This requires a level of output in each that reduces costs to international levels.* Structuring specialization in this way would require an intergovernmental agreement. Such an agreement would be conceivable *within* the Latin American countries and might become a base for an extension of the same type of arrangement as exists between the United States and Canada, in which the Canadian subsidiary has been assigned a given line of vehicles by the parent company. But the decision on the form of specialization has been made by the international companies.

If Latin American countries cannot get together and make such an arrangement with the United States collectively, Mexico or Brazil might "integrate" with the United States on their own. During interviews in Latin America, several officials in Brazil, Mexico, and Argentina stated that it would not be to their country's interest to integrate with other Latin American countries. If they did integrate the industry with another country, preferably it would be with the United States. Given the existence of in-

---

* The necessity for large-scale output is tied to the objective of achieving international levels of costs in order to sell on the world market. If the only market to be served is the national market (or possibly some other less-developed country), it is quite possible that a jeep-type vehicle convertible to passenger, taxi, or small truck operation could be produced quite efficiently. Costs lower than those now in the production of more sophisticated models could certainly be reached, especially in countries where labor costs remain low. Thus, production of vehicles other than those usually designed by the international companies is feasible for markets not demanding more sophisticated vehicles.

Ford has recently decided that there is such a differentiated market in Southeast Asia and is proceeding to establish plants in several countries that would produce components for a very simple vehicle that could be assembled in several countries for the local market.

The alternatives examined in this paper are, therefore, not all those that are available.

ternational companies, such an integration is not uneconomic; it merely seems unfeasible given the absence of the same kinds of ties that exist between the United States and Canada. It is also likely that any attempt by a single country to integrate with the United States would be strongly opposed by other Latin American countries, since this would reduce the attraction of regional integration and alter the industrial balance among these countries significantly.

Regional Integration. Full LAFTA integration could achieve international competitiveness for the industry.* This could be accomplished by mere reduction of internal barriers, letting competition eliminate the inefficient producers and relocate production facilities, or by forcing a restructuring of the industry under selected companies. The companies selected could be given the Latin American market for particular types of vehicles or models, thus expanding their scale of operation. Or the national companies selected could produce various components, exchanging them and assembling in the national market, which would be largely reserved for them. Alternatively, several companies could be selected by all Latin American governments to produce throughout the region. Repartition of components, models, or types of vehicles would be made under intergovernmental agreement. The various impacts of and obstacles to these alternatives are the subject of the succeeding chapter.

Subregional Groupings. If any of the major countries opts out of a plan for regional integration, the remainder have the same choices as above, though the level of cost reduction and efficiency will depend heavily on the remaining size of the market and the number of models or types of vehicles agreed upon.

If the three major countries opt out of integration schemes, the Andean group remains, with or without Venezuela. With Venezuela, the total market by 1980 would amount to nearly 500,000 units—enough for a single strong company, or a few companies producing distinctly different models. Without Venezuela, the Andean market by 1980 would amount to something over 300,000 units. At levels of 70 per cent local (Andean) content, costs would still be substantially above international levels but significantly lower than existing costs in these same countries. To achieve significant reductions would require rationalization of the industry into one or two companies producing a few models. The duplication arising from the existence of six to ten companies in each national economy would have to be eliminated. The same obstacles arise with full LAFTA regional integration, though in a more acute form; this is discussed in succeeding chapters.

---

* See Appendix A.

National Rationalization. In the absence of moves to regional integration, the remaining alternative is a restructuring of the national industry. Peru, Chile, and Argentina have moved in this direction by substantially different routes. Peru has recently required the elimination of several of the assemblage companies, reducing the producers to four in number; it is likely that it will move to consolidate these still further and has suggested that it may follow Chile's lead. Chile has begun discussions with a Japanese company not now counted among its nine producing companies with the objective of forming a monopoly, with majority ownership by the government which would produce a limited number of models at low cost; it would probably take over gradually the operations of the existing private companies. Argentina has begun to put price and cost pressure on the existing companies and their suppliers by letting more imports into the country; it can use this tactic so long as its foreign-exchange reserves are adequate. If they are not sufficient, other tactics will probably be used to restructure the industry, which has been reduced from twenty-two companies to less than ten and will probably shrink to less than a half-dozen under increasing governmental pressure.

The visibility of national rationalization depends on the growth of the market; national markets will remain rather small in all but the two largest countries, Brazil and Mexico. Argentina can look forward to a market of 400,000 to 600,000 units by 1980, but this means only one company with a few models, or a few companies with one or two models each. Such an approach is certainly feasible and is made even more economic if local-content requirements are not stringent; more companies and more models would then be feasible. But the Andean countries cannot look forward to any significant national markets. The best that each can do to cut costs to reasonable levels would be to create a single company, producing one model. The cost of this policy is not borne so much in high prices as in a restricted choice for consumers—a cost readily paid by socialist-oriented governments.

National rationalization can make the industry more efficient, but it also reduces the attractiveness of regional integration. It reduces the differential benefits and increases the obstacles by making the national structure more rigid and injecting state-owned enterprises into the structure. It is difficult to integrate operations that are vertically integrated and have become rigidly tied to a government-controlled market. And it is exceedingly difficult to make the inter-enterprise adjustments needed to merge state-owned enterprises. The interests of ministries and political pressures are added to the normal difficulties of transnational mergers. In addition, the existence of military governments (as in Peru) injects still more strongly the national security issues that are often just below the surface of discussions of integration. The longer that moves toward integration are delayed, the less likely is the success of any future move to industrial integration in the auto-motive industry. Having achieved a foothold, governments are likely to proceed to *national* integration and are not likely to sit by waiting for some-one else to determine how to make the industry more efficient or useful.

The Petrochemical Industry

As in the case of automobiles, governmental policies have induced duplica-tion of production facilities in petrochemicals; but the structure of owner-ship, vertical integration, trade, and prices differs sharply from that in the auto industry. Again, though there would be substantial benefits from inte-gration, it is difficult to determine which routes would provide the best balance among the many trade-offs required.*

Mexico stands first in present capacity in petrochemicals. Because it started earlier, it has achieved a more-developed market, more-advanced technologies, and lower-cost inputs than have the other countries. Even so, it produces only three-fourths of its own consumption and is not a major producer compared to advanced countries. In an effort to build a petro-chemical complex, the government has stimulated production by Petróleos Mexicanos (PEMEX) and induced foreign investment, leading to over-capacity in many lines.

Brazil is moving toward the same goal by creating a strong national in-dustry and has recently embarked on a large-scale petroleum complex. Nevertheless, it still must import petroleum and also lacks sulphur. Its major assets are low-cost labor and a rapidly expanding market. The items into which it is moving track almost exactly with those produced in Mexico.

Argentina is relatively well endowed with raw materials inputs for petro-chemical production, but it lacks sulphur and has relatively high-cost labor. By moving strongly into intermediate products, it is trying to regain a pri-mary position, which it lost to Mexico some years back.

Venezuela's resources in petroleum make it one of the largest producers in the world, but the lack of a rapidly expanding domestic market for chemicals has meant that petrochemical production has not been stimulated. The combined output of the chemical industries in Venezuela, Colombia, Peru, and Chile is still smaller than that in Argentina, which is third behind Mexico and Brazil.

Each of the major countries has established a government-owned com-pany to exploit petroleum resources and develop basic petrochemical pro-duction: PEMEX in Mexico, PETROBRAS in Brazil, Yacimientos Petrolíferos Fiscales (YPF) in Argentina, Instituto Venezolano de Petroquímica (IVP)

---

* See Appendix B.

in Venezuela, and Empresa Colombiana de Petróleos (ECOPETROL) in Colombia. The state-owned enterprises in the big three have a monopoly in processing crude petroleum; many companies produce intermediates, and a few have established production of some final products. In Mexico, foreign-owned enterprises are being pressed to sell a majority ownership to local interests. But Brazil and Argentina are rather liberal as to foreign ownership of petrochemical facilities. The Argentine government has entered a joint venture with Dow Chemical Company for the production of several important intermediates; this joint venture pattern between state-owned and foreign enterprises is also found in Mexico, Venezuela, Brazil, Chile, and Colombia.

The international companies which have been induced to enter the region have brought important technological advances and provide the continuing inflow of technology needed to keep the industry up to date. Some eighteen to twenty United States and European companies have affiliates in Latin America, producing intermediates and final products. They are oriented almost wholly to the domestic market, as a consequence of governmental restrictions.

Because of the government ownership of production of basic chemicals and some intermediates, it is not possible for the international companies to integrate backwards toward basic materials. They can move forward from intermediates to final products, but this ability depends on the size and growth of the market, which has been expanding rapidly in only a few countries. Still, working through joint ventures, each of the major countries is seeking to develop a vertically integrated industry prior to negotiating any serious reduction of trade barriers.

Intraregional trade in petrochemicals has been greatly restricted by both tariffs and quotas, with permissions to import dependent on whether a local supply exists or is in the offing. The opposite side of the coin is that Latin American petrochemical exports are low. Mexico's exports of all chemical production amounted to only 4 per cent in 1967, with only about 40 per cent of this being petrochemicals. Of the total petrochemicals exported by Mexico, less than one-third went to other Latin American countries. Exports from Brazil also have been virtually nonexistent, and Argentina has priced itself out of exports.

Price levels in petrochemicals are influenced by the restriction of imports, small-scale operations, and government ownership of basic chemical production. Mexican prices, for example, are reportedly 30 to 40 per cent above international levels for intermediates, and final-product prices have been found to range from 40 to 115 per cent above U.S. prices of the same items, partly because of the high prices of intermediates. Brazil's prices are strongly influenced by the high cost of transport of imported raw materials, plus high capital costs. Its prices range from 20 to 160 per cent higher than international levels, with the exception of SBR rubber, which it can sell below international prices. Argentina's price structure is also heavily influenced by the costs of basic chemicals, which in turn is a function of small-scale operations; prices range from 45 to 145 per cent above international levels.

Benefits of Integration

Given the small size of national markets for petrochemicals and the potential economies of scale from integration, there are substantial benefits to be gained from restructuring the industry to serve the LAFTA region. Efficiencies would arise from a more continuous use of equipment and manpower, reducing the effects of overcapacity, and from rationalizing the sources of supply for markets.

Economies of scale would result from improved investment programs, from reduction in costs of inputs, and from more efficient operation of facilities. Unit costs, for example, drop precipitously as plant capacity increases in many intermediates: it takes only twice as much labor to operate a plant with capacity ten times larger. Supervisory staff remains fixed over a wide range of output. Cost of materials inputs should be stabilized and reduced by integration, for the impacts of national policies on state-owned companies would be mitigated.

Plants of adequately efficient scale for basic and many intermediates would be supported by an integrated market made up of the combined demand of the three major countries by themselves; aggregate demand of the entire region would provide a growth factor that would permit still larger operations using technologies which may be developed in the future. The location of the least-cost facilities would tend, however, to be in the large-size countries (though they are not the least-cost producers) simply because they have the resources to build an entire petrochemical complex. Colombia happens to be the least-cost producer of a wide range of petrochemicals, but it could not commit sufficient resources to produce all of them. Colombia therefore would have to specialize, leaving many items to the larger countries.

Final products do not require large-scale operations and can be produced economically in several of the Latin American countries. The economies tend to arise from being close to the market. Again, since the markets are the largest in the large countries, the benefits of integration will tend to flow to them. Mexico is likely to gain the most, not only from final products but also from the fact that it is the second lowest-cost producer in intermediates.

Alternative Patterns

As suggested above, many of the gains of integration can be achieved by rationalization of the industry among Mexico, Brazil, and Argentina, the three countries having the greatest petrochemical capacity. As noted earlier, these three have recently joined in a complementation agreement with Venezuela, whereby they will reduce duties on 300 items, though they made only 67 concessions initially (the fifteenth agreement). There is also the Andean group complementation agreement in petrochemicals (the sixth), but it appears to have much less viability. The group has agreed to reduce duties but also to allocate production of all petrochemicals within the region.

Among the four major producers, Mexico is likely to be the least-cost producer of basic petrochemicals, such as ethylene, benzene, and orthoxylene. Since these chemicals are costly to transport, a complex of intermediate plants is likely to grow around the Mexican production if duties are removed on these items. For example, Mexico would likely be the least-cost producer of styrene and polystyrene.

A conflict arises in the determination of the location of production. If one works backward, the three major countries (excluding Venezuela) will probably have final-product demand that will support large-scale (efficient) production of intermediate monomers. But to have to obtain the basic materials from Mexico would raise the costs. A trade-off arises for the three countries between high transport costs of basic chemicals and high production costs of smaller-scale production of the same items locally.

Argentina's advantage lies in its low-cost natural gas, which would support efficient production of fertilizers. But its national demand is not sufficient to absorb large amounts, and costs of transport are high. Argentina might be left with production of ammonia for fertilizer, which would be shipped to Brazil for processing into ammonium nitrate or sulphate. Mexico could produce its own fertilizers. Other products that Argentina might produce on a least-cost basis do not have a sufficient market in the three countries combined; one such product is methanol. Argentina's gains must wait on a substantial increase in incomes to raise demand for final products.

Brazil's advantage would lie in producing items for heavy industry close to the market, e.g., the automotive industry or those industries using polyester fibers. But its total output would still be smaller than that going to Mexico.

Venezuela's position in the new arrangement is uncertain. It can produce ethylene cheaply, but it is equally costly to ship that intermediate to Brazil's industrial centers and to Argentina as it is from Mexico. All four of the countries that signed the agreement are seeking to develop an entire petrochemical complex of the most modern sort, and the three South American countries are drawing in the international companies to make production of intermediates more efficient. It is not yet clear what production—if any —they have given up in agreement.

Integration among the four countries would still leave a cost gap compared to international levels ranging from 30 to 45 per cent. But the cost reductions gained from their integration under free trade would range from 3 to 95 per cent over a selected list of products.

Gains in terms of balance of payments have been estimated (looking only at the potential changes in the sector) at $20 to $40 million through import substitution vis-à-vis outside countries. A new pattern of trade would emerge among the three big countries. Mexico would be the biggest gainer by far, winding up with a net surplus, while Argentina and Brazil would be left in a net deficit position with the outside world. However, these deficits would be significantly reduced or even eliminated by their new ability to sell to third countries; and Mexico would gain still more through third-country exports. Mexico could gain something like $125 million in its payments position, while Brazil and Argentina might end up with merely a balance in petrochemical trade.

In terms of the use of advanced technologies, Mexico would tend to gain more than the others, though Argentina would gain more than Brazil. In terms of employment, the shifts in petrochemicals are not likely to cause significant gains, but former operations using nonpetrochemical techniques would decline, adding to unemployment pressures. The slack would be taken up temporarily through construction of new plants and later by expansion of final-products output.

The Andean group includes Colombia, with petroleum; Chile and Peru, with abundant cupric, nitric, and phosphate resources; and Ecuador and Bolivia, which have few resources. Under free trade, Colombia would be the least-cost producer in the group. Peru would produce fertilizer for all. Chile would tend to be left out under free trade, as would Bolivia and Ecuador.

Just as most production would shift to Colombia, so would most of the benefits in terms of balance of payments, technology, employment, and development of secondary industry. This potential result undoubtedly was a strong reason why Venezuela refused to join the Andean countries and why they allocated production—though no facilities have yet been erected under the terms of the agreement.

If the entire petrochemical markets and industries of LAFTA were integrated, merging the two groups, there would be a still further shift in the location of production under free trade. Colombia would tend to be the least-cost producer for practically every major item. But would it then become the center for the petrochemical industry for all of Latin America? As previously observed, it does not have within itself adequate resources to produce large volumes of the entire range of chemicals; furthermore, its location is far from the major markets—Brazil's cities, Argentina, and Mexico.

The location of these markets indicates that a substantial amount of production of intermediates should take place within the three larger countries, even if Colombia were the center and produced, for example, all of the basic chemical ethylene. It would be more economic to place production of styrene, adipic acid, butadiene, and other intermediates closer to the final market. Colombia, however, would displace Mexico's production of ethylene and shift the economic base for production of intermediates out of Mexico to the other two major countries as well. The location of fertilizer demand would indicate that, as under the subregional plans, it should be produced in Mexico, Brazil, and Peru. Likewise, plastics and synthetic fibers would tend, as under the subregional plans, to be produced near the final markets.

The fundamental difference made by full LAFTA integration is to shift the locus of basic chemical production and some of the intermediates out of Mexico into Colombia and the two other major countries. For Argentina and Brazil, the shift is from sourcing of basic chemicals in Mexico to sourcing in Colombia. Production of SBR rubber would shift similarly. But no significant shifts would occur within the Andean region as compared to that under free trade within it. However, the Andean group would not accept the results of integration under subregional free trade, for the members could not accept the resulting allocation of production. And substantial additional changes would have to be made in the allocation of production if it merged with the big four. Such a merger would be made more difficult because of the necessity to balance the distribution of the benefits. Substantial gains from LAFTA integration would not be achieved for the eastern region (under free trade), and the shift out of Mexico to Colombia would redistribute some gains outside the three major countries. The western group would gain substantially by greater increases in efficiency, employment, trade, and technology. Some means would have to be found of redressing the balance.

The gains in efficiency from combining would be significant. Producers in the Andean group could drop average production costs of a range of ten selected chemicals from 20 per cent above U.S. levels to merely 2 per cent above U.S. levels; costs of those fewer chemicals likely to be produced in the subregion could be cut to levels equal or less than average U.S. levels. Three new products could be produced efficiently: methyl metacrylate, melamine, and dimethylterephthalate, but not at levels fully commensurate with those in the United States. In any event, imports of these items were only $5 million in 1966, making for no great gain from import substitution by LAFTA production. The addition of the Andean demand does permit duplication of production facilities for a few of the chemicals already being produced in one subregion, so that the benefits of employment could be spread with no loss to efficiency. However, in a large portion of the chemicals, the combined regional demand still would not permit erection of plants of the same scale as those in advanced countries. Even if the resulting units are adequately efficient to sell products on the world market, they are not likely to be competitive with the most efficient plants in advanced industries.

In sum, full LAFTA integration would not increase the efficiency of the plants of the four major producers significantly over their subregional integration but it would benefit those in the Andean region.

More significant changes would occur in the structure of trade under LAFTA integration as compared to subregional groupings. Andean countries would substitute eastern production for products imported from outside the region. There would be some regionwide import substitution for outside products. And the big four would shift some purchases from Mexico to Colombia. The first two impacts would save foreign exchange for the region as a whole, though not a large amount compared to the prior savings under the two subregional groupings. The total savings would amount to about $27 million—$22 million for shifts from extraregional to regional purchases by the Andean subregion and $5 million from new production of import substituting items. The latter amount would be produced in Colombia, so trade and employment benefits would accrue to it. Colombia would also gain from a shift of some exports from Mexico.

Employment gains to the region as a whole would arise from new import substitution amounting to roughly $27 million. Since the average output per worker in the industry is $25,000, the employment gains would amount to about 1,000 workers—about 400 of these in Colombia, with the rest spread over the other countries. Shifts in employment would also occur between Mexico and Colombia and an increase in employment in the industries of the east as they expanded to meet the new western demand.

Practically no net gain would be generated for the full region in use of new technologies, for the most-advanced technologies would already be used in subregional integration. Only in the introduction of a few new, import-substituting products would additional advanced technology be acquired. However, the shift of the petrochemical center to Colombia would mean that much of the advanced technology would flow there, rather than to Mexico.

Comparison of Alternative Groupings

Integration among the four major producers would achieve almost all of the gains for them that would also arise out of LAFTA integration, with the offsetting result that Mexico would lose some exports, technology, and employment if the Andean region were included. The fact that full LAFTA integration would add no new products to the industries of the four biggest producers means that they can gain only in increased efficiency and employment within industries that would exist under subregional integration. The major *gains* in efficiency, employment, trade, and technology would go to the Andean countries, which are the relatively less advanced. But in the process, control over the center of much petrochemical activity would pass to Colombia under free trade.

The justification for full LAFTA integration as compared to subgroupings will, therefore, likely stem from *noneconomic* arguments concerning the equitable distribution of the benefits. If the big four are persuaded that it is good for Latin America or themselves that the Andean countries get an industrial stimulus through petrochemical development, integration might be acceptable. Without such a view, either integration will not occur *or* it will not occur through the simple removal of barriers. The fact that benefits can be gained is not a sufficient pressure given the uncertainty or dislike of the potential *distribution* of those benefits. Though integration theory indicates that substantial benefits would arise for the region as a whole as a result of integration, what national governments and industries will be concerned with is *how* those benefits will be distributed.* And they will want to know this beforehand—*not* wait until all benefits are known and then have to find a way to redistribute the additional income.

Both subregional integration and full LAFTA integration raise the problem of the distribution of the benefits and the trade-offs of one against another, especially against that of efficiency and least-cost solutions. Only if the governments are individually and collectively concerned with the maximum benefits to the region as a whole will they accede to LAFTA or subregional integration without making significant changes in the results projected above. It is the essence of national policy, however, to look after the national benefit. Governments will, therefore, have to be satisfied that either the process of deciding where the activity will be located is likely to produce equitable results, or the agreement to integrate must spell out an equitable structure. Either requires trade-offs between efficiency and equity.

---

* For this theory and models, see Appendix B.

The Trade-offs: Efficiency Versus Equity

In both the automotive and chemical industries, an integrated Latin America could reach economies of scale that would cut costs to levels equal to those in the advanced countries. Both industries would probably find it possible to export some of their production, contributing to foreign-exchange earnings. This goal could be achieved, technically and economically, within ten years. Part of this time would be needed merely to wait for demand to rise to levels that will enable appropriate economies of scale to be reached. Simultaneously, the new programs of investment and industry relocation would have to go forward. This is not to say, however, that the technical and economic resources will be committed to the task. There are available various approaches embodying various technical, economic, and commercial combinations that would produce approximately the same efficiency levels in the two industries. But the several approaches would distribute the benefits quite differently among the participating countries. In chemicals, Mexico would be a substantial gainer in a subregional group with Brazil and Argentina, but Colombia would gain still more under full LAFTA integration.* In automobiles, Mexico and Brazil would be substantial gainers if the markets of the entire region were opened to all; if they were not opened, these two countries may achieve different levels on their own and bypass the rest of the region.

If each country is to benefit somewhat equally under full LAFTA integration, the increase in efficiency will be less. The tendency will be to move away from the most efficient arrangement toward one in which benefits are seen as equitable. The resulting trade-offs between efficiency and equity would arise from decisions as to (1) the extent of local production versus imported components and (2) the extent and nature of specialization among the member countries.

The objective, of course, would be to gain international levels of efficiency while permitting all countries to participate in acceptable ways and at acceptable levels—"acceptable" in the sense that they agree to participate and seek an efficient operation. Giving prominence to equity considerations raises the questions of how to meld the national suppliers in the auto industry, the state enterprises in petrochemicals, and the international companies in both industries. National suppliers cannot be sacrificed for political reasons; state enterprises are likely to be seen as means of re-

---

* See Appendix B, "Alternative Patterns for Integration."

49

taining national bargaining power; and the international companies will be needed to make integration more effective and to maintain advanced levels of technology.

Local Content Versus Exchange Costs

Lower cost-levels can be obtained in each industry by import substitution on a regional basis. This could be achieved by employing the concept of regional local content and permitting products to move freely with no national trade barriers. But the shift from national to regional content raises the problem of equitable distribution of production facilities among the countries. Regional import substitution shifts trade patterns in such a way that foreign-exchange costs and earnings from *outside* the region are partly replaced by foreign-exchange costs and earnings *inside* the region. The distribution of balance-of-payments gains and losses becomes a critical issue of equity.

The concern of governments for foreign-exchange costs is shown in the high barriers to imports of automobiles: from 70 per cent in Peru to 305 per cent in Chile; a low 12 per cent duty in Argentina is made irrelevant by a license requirement limiting the number imported. In other countries, the burden of high duties is raised still higher by prior deposit requirements and surcharges. Colombia permits a large number of fully assembled vehicles to enter the country, but it does so under bilateral trade deals (e.g., the arrangement with Czechoslovakia, whereby Skoda automobiles are exchanged for coffee outside of the quotas established by the international coffee agreement). Given the concern over payments deficits, it may be assumed that governments will be reluctant to enter integration schemes in the auto industry that would involve any greater demand on their exchange earnings (or at least any greater *proportion* of exchange earnings) than has been taken by imports in the past. A net-exchange drain to another Latin American country is no better than one to a country outside the region, if payment has to be made in convertible currencies.

An example of the choices required is seen in the alternative routes to development of the automotive industry of the Andean countries. Assuming that the amount of vehicles purchased would be the same over any range of prices and that each country would impose the same level of local content, the purchases in 1975 by the group would aggregate about 140,000 vehicles. At present international prices, this would amount to $450 million of outside foreign exchange if all were imported. The following table shows the comparative costs of 140,000 cars, estimated for different levels of national and regional manufacture:

Production costs of 140,000 vehicles		Outside foreign-exchange costs	
		(millions)	
		Direct	*Indirect*
Imported	$450	$450	—
National content:			
100%	925	—	$150
50%	580	290	30
Regional content:			
100%	590	—	50
50%	500	250	25

Given national production and local-content requirements of 100 per cent, these countries could, among themselves and separately, produce the same number of vehicles at a total cost of $475 million *above* international prices, with outside foreign-exchange costs for materials no greater than $150 million. At 50 per cent national content, the cost penalty would be only $130 million, and outside exchange costs would total over $320 million, of which $290 million would be direct for automobile components and $30 million for materials from outside the region. But, if these countries integrated their industries and substituted a 100 per cent regional-content requirement, our calculations indicate that the cost penalty for the same number of vehicles would drop to $140 million and at 50 per cent regional content to only $50 million. Regional integration could save $335 million compared to national production costs at 100 per cent content and $425 million at 50 per cent content. Outside foreign-exchange costs could be cut from $150 million at 100 per cent national content to $50 million at 100 per cent regional content. Moving from 50 per cent national content to 50 per cent regional content saves only another $45 million in foreign exchange. The major savings for moving to regional integration would be in internal costs, compared to outside foreign-exchange savings. The optimum situation from the standpoint of both kinds of cost is regional integration with 100 per cent local content required.

But these calculations say nothing of the distribution of *internal* foreign-exchange earnings or costs as members trade among themselves. (Nor does it say anything about the number of companies or their ownership.) For some countries, *inside* foreign-exchange costs would exceed earnings; for others, the opposite would be true. Whether the Andean countries will actually integrate may be expected to depend on the distribution among them of the various savings and costs.

If all LAFTA countries were included in the integration scheme, total demand would be ten times that of the Andean countries—1.4 million units—permitting vertical elimination of cost penalties against international levels, assuming the industry was restructured so that each company produced over 400,000 units annually. There would be no need to trade off cost increases against outside balance-of-payments savings; whatever is saved in outside foreign exchange is all to the good. But, again, the exchange savings would not necessarily be distributed equitably among the participating countries.

It is of little advantage to Peru to substitute exchange payments of $3,000 per vehicle to the three big Latin American countries for similar exchange payments to the United States. It is even less attractive to substitute payments of $4,000 to other Andean countries for the $3,000 formerly paid to the United States. To induce Peru to join the Andean group at a cost penalty of $1,000, it would *at least* have to obtain production or exports equal to $1,000, so as to leave it no worse off in net foreign-exchange payments. But something more would likely be needed. For example, Peru would probably demand that at least 30 per cent of the value of the Andean vehicle be made in Peru—in other words, a national content of 30 per cent. Or it might insist on exporting an equal amount for any drop below the 30 per cent level. If it had only a 20 per cent local content, it would have to export an amount equal to half its local production to other partners in order to meet the extra exchange cost.

By way of illustration, assume that each of four countries in the Andean group, prior to integration, is importing all vehicles from the United States at an average cost of $3,000. With integration, the average cost will rise to $4,000 at 100 per cent regional content. To produce four models, so that each country has one, the four countries will have to produce $16,000 worth of matched components. The following distribution of production and trade might obtain under free trade:

	Production	Exports	Imports	Consumption*	Exchange Position
Peru	$4,000	$2,000	$2,000	$4,000	—
Chile	2,000	1,500	3,500	4,000	− $2,000
Colombia	9,000	5,000	—	4,000	+ 5,000
Ecuador	1,000	750	3,750	4,000	− 3,000

* Cost of one vehicle figured at $4,000.

The $12,000 of foreign exchange savings are distributed quite unequally: Ecuador has no net change though its imports are shifted from the United States to within the region; Colombia gains $5,000 in new exports plus recouping the $3,000 formerly spent, with a net of $8,000; Peru gains from import substitution by $3,000 and Chile gains by $1,000.

Integration would supposedly eliminate the former prohibitions on auto imports. The amount demanded would probably rise, even with a price increase; total exchange costs would tend to rise above the former payments to the United States. The limit a country might place on its exchange costs would probably not be per vehicle as in the above model but on aggregate payments. Thus, if Chile previously imported 100 cars at $3,000 each, its exchange limit would be $300,000, which could be spread over 150 vehicles if the average exchange cost were cut to $2,000 by integration.

Although the above distribution of production within the region might provide the least-cost vehicle, it is doubtful that the members would accept this distribution of gains in terms of exchange. They are free, of course, to alter it by allocating production differently, but they would do so at a further cost penalty. They can reduce this internal cost penalty by relaxing the content requirement. For example, if they dropped to 50 per cent regional content, they would have only about a 10 per cent cost penalty to absorb, making the price of the vehicle $3,300. But they have only half of its value to divide among the local producers. To achieve equity at a lower cost-penalty, there is less local production to divide and foreign-exchange costs rise severalfold, though still smaller than under unrestricted imports or *national* content.

Alternatively, a low cost-penalty could be obtained with a more equitable division of exchange costs and gains by widening the group integrating. Even at a 100 per cent regional content, the entire LAFTA group could be organized and produce at levels of efficiency involving no cost-penalty. There may even be some leeway in the allocation of production and structure of intraregional trade before cost penalties are incurred. But the achievement of equity does involve a repartition of the location of production among the countries so as to balance imports and exports appropriately.

The above example is in the automotive industry, but exactly the same trade-offs would arise in the petrochemical industry. Any country can choose to import basic, intermediate, or final chemicals from outside the region; to import from within the region; or to produce within the country. A balance between total cost and exchange cost must be struck that is acceptable.

Pattern of Specialization

There are at least four kinds of specialization that government officials seem to be taking into account in determining whether the location of auto production would be equitable: technological specialization, component specialization, model or type-of-vehicle specialization, and company specialization. The production pattern required to make foreign exchange gains equitable will not necessarily accord with any of the criteria of specialization. It is possible to combine these types of specialization in several ways to obtain different impacts; trade-offs among them will be necessary in order to help balance the distribution of other benefits discussed in this chapter. In addition, any selected combination of these four types of specialization is likely to require a trade-off against specialization for efficiency, i.e., what would be dictated by a free market under free trade. However, since it is evident that this "efficiency" specialization is not acceptable to governments, even if attainable, the question is how much are they willing to pay for any of the other (more equitable) forms of specialization.

Technological specialization concerns the distribution of advanced technologies among member countries. In motor vehicles, only a few trade-offs in the technological area are available, for the most advanced techniques are in engines and transmissions, followed by axles and stampings, with little else that matches these levels save in some accessories. The most efficient level of output for engines is 500,000—larger than any one company is likely to be through integration even by 1980. Even if all the engine producers were separated from other companies, there would be only three of them. Where should they be located? To insist on local production within the major producers, as is the case now, would be to continue to pay a high cost-penalty. To integrate but permit engine production only where it is currently, would raise the cost penalty for those countries (e.g., the Andean group) that now import their engines from outside the region.

In the petrochemical industry, efficiency requires that the more advanced technologies be aggregated in one complex. To redistribute production and the use of those technologies among countries is to raise costs substantially. But no major country would be willing to permit such advanced technologies to be concentrated in another country in Latin America. Rather, each is seeking to gain a foothold in the advanced technologies in order to gain more under integration.

Strategic interests are overlaid on the technology issue. The military of each major country consider it a necessity to be able to provide essential components and parts of its own motor vehicles. For example, neither Argentina nor Brazil would permit the other to specialize in automotive engines if it meant the loss of the ability to produce military truck and vehicle engines within that country. To solve the technological specialization problem, it would be feasible to permit specialization according to engine types. One country would produce truck engines; another four-cylinder engines; another sixes (there might be enough sixes to permit duplication); and another eights. Conceivably there could be five or six lines producing engines. It is not likely that this type of division would be acceptable either, but it would share the technology. Given an output of 3 million units in the 1980's, the same type of division would be made for transmissions; again, each military establishment would probably want the automatic transmissions located in its country.

Specialization according to components could expand the economies of scale in each major subassembly. Given that the economies of scale in body stampings are reached only at a level of 1 million, by 1980 it would still be most efficient to produce them in only three locations in Latin America. But this would give all of the steel demand to three countries, unless sales of hot-rolled sheet steel were allocated among other producers also. Either Brazil or Mexico could come close to having an efficient scale for stampings by that time from internal demand, but only one company would exist in each. Who would get the third plant?

Specialization by component is relatively easy in the auto industry because of the large number of standardized parts which can be readily produced in many locations: radiators, upholstery, glass, headlights, batteries, tires, wheels, front axles, brake linings, cables, instruments, radios, heaters, air conditioners, and so forth. They do not require the same scale and are easily allocated among countries to achieve balance in production.

Similar specialization is not as easy in the petrochemical industry. Some products have to be produced near the basic or intermediate chemicals or near the market in order to achieve efficiency. Specialization by basic products or intermediates leads to cost penalties; rather than making trade-offs easy, this makes them more difficult. Integration backward and forward is more at a premium in petrochemicals than in the auto industry. The difficulties are illustrated in the Andean complementation agreement on petrochemicals, which allocated production of specific chemicals to the member countries in order to spread the benefits equitably. Bolivia, for example, was assigned styrene and hexamethylene diamene among other products; on the basis of the cost structure of the four countries involved, this location will involve substantial cost penalties. Styrene is important in a wide range of products, especially polystyrene and SBR rubber; and hexamethylene diamene is a basic raw material in nylon 6–6; yet neither of these basic chemicals can be economically produced in or transported from Bolivia. Polyvinyl chloride was allocated to Peru and vinyl acetate to Chile, but in neither case is the assigned country the least-cost producer of the chemical. A basic decision, therefore, was made in favor of equity over efficiency. To achieve this pattern, a technical commission is provided for under the agreement, with the power to close plants and alter the product mix of remaining ones.

A third type of specialization is available in the automotive industry— by models or types of vehicles—but not in petrochemicals, apart from product lines. Each country could produce one model or type of vehicle, restricting itself to final assembly and specialized components. For example, Mexico might concentrate on larger passenger vehicles, Brazil on small cars and Argentina on intermediate-sized cars with other countries specializing in commercial trucks, heavy trucks, trailer tractors, or buses. Economies of scale would exist at the stages of final assembly, subassembly, casting, stampings, engines, and transmissions, for each of these would be substantially different in each country. Some of the international companies have proposed just such a specialization within their own operations throughout Latin America—so far to no avail.

Product specialization could be employed in the petrochemical industry: Peru would concentrate in fertilizers, Brazil in synthetic rubbers, Colombia in ethylene, another country in plasticizers, and so on. This type of specialization is not contrary to efficiency criteria if confined largely to end products, for little efficiency is gained from wide product diversification within one country. Efficiency is largely gained from vertical integration within an economic region. Product specialization would lead to inefficiencies only if it separated (uneconomically) the intermediates from basic chemicals.

Finally, specialization could be permitted on the basis of automobile companies. Rather than having each company spreading its various production activities over several countries, and trying to bring together the various components for assembly where it finds its markets, each country could contain one or more companies—just large enough to supply its market at competitive prices. For example, FIAT might take the Argentine automobile market, VW and GM the Brazilian market, Ford and GM the Mexican market, and Chrysler the Venezuelan market; other companies might combine or opt out, leaving only one or two to serve the Andean group. It could then be left to these companies to organize their supplier relationships to make their operations the most efficient possible. Similarly, the petrochemical industry could be organized according to major companies—either the state enterprises or the international companies or some new type of Latin American multinational enterprise.

The gains for each country from each type of specialization would be quite different, and the impacts on existing operations would be substantially different. The balancing of the various impacts would require a more thorough analysis of the existing and potential relationships than has been made for the present study; such an analysis would be required in order to provide sufficient information for policy decisions of governments. Even without a complete examination of the various combinations that could be devised to maximize equity with a minimum sacrifice of efficiency, it is still possible to denote at least three areas in which further trade-offs will be required and which would cause difficult problems for governments. They relate to the interests and best utilization of existing organizations: the state enterprises in the petrochemical industry, the local suppliers in the auto industry, and the international companies in each.

Auto Suppliers

The major means that host governments have used to inject local ownership in the auto industry, while welcoming the international companies, has been to reserve the supplier segment for local investors. Mexico has required majority domestic ownership of suppliers and enforced this by requiring prior approval of investment by foreigners. Colombia, Chile, and Peru also have local-ownership requirements, and Argentina is moving in this direction. These local-ownership policies would be threatened by integration if the assembling companies were permitted to manufacture components in one country and export to its affiliates in others or if mergers were encouraged among supplier companies across national boundaries. The latter process would accord with "Latin Americanization," but the former would leave the foreigner in control.

The concern to protect local suppliers and to minimize exchange costs is illustrated by the present requirements that an assembler export components in order to receive permission to import components similar to those already produced within the host country. In Mexico, for example, Nissan was allowed to import radios from Chile *if* it maintained purchase of local radios at the previous levels *and* if it exported locally made engines to a third country for an equal amount of exchange. Automex (Chrysler) was allowed to make a similar swap but was also permitted to reduce its domestic purchases of the parts it was importing by 10 per cent per year. In Argentina, the 1965 law allowing imports to be counted as local content is restricted so that an equal value must be exported and the amount imported cannot be more than 30 per cent of the assembler's requirements of that particular item.

The problem of how to protect the interests of existing local suppliers is so acute that the Mexican secretary of industry and commerce reportedly stated that their existence precluded any Mexican interest in complementation agreements in automobiles, except in the most marginal of areas. Complicating the problem of local suppliers is the fact that by no means are all of them locally owned. Some of the suppliers of key components are affiliates of U.S. companies; some of these are jointly owned with local investors. And some of the assembling companies are integrated backward into production of the major and minor components, starting with engines and moving into more minute parts.

The assemblers that also produce components have considered regional integration wholly desirable; the increase of specialization and potential for swapping components among affiliates would reduce costs and stimulate sales, leading to further expansion of output and cost reduction. Some independent suppliers of components are in a similar position, especially those producing items that require large amounts of specialized equipment and high technology. Intraregional trade in these items would achieve economy of scale (examples include transmissions, gear boxes, steering assemblies, rear axles, and radios). Most of these firms are U.S.-owned and have followed the assemblers into the countries where final products are made. They are usually the sole or principal suppliers and have a dominant position throughout Latin America; they have little to fear from competition throughout the region. Some competition might arise from assemblers themselves, as scale advantages increased, but it is likely that the U.S. pattern of relying on independent specialized companies would be followed. The position of the independent suppliers is strengthened by their ability to serve several assemblers at the same time.

The third type of supplier is usually locally owned, small, and reliant on low-level technology. However, some of these local companies have strong positions as sole suppliers of key components. For some components there are several local sources. Other local suppliers face competition from international companies, such as in the case of tires and paints. In some instances, the locally owned supplier is dependent on technology or equipment supplied by the assembler. And in still others, the assembler holds a portion of the equity shares of the supplier; the share tends to be larger the more advanced the technology that is transferred.

Governments have supported the supplier companies by exempting them from the price controls imposed on vehicles (Mexico has frozen prices of automobiles since 1966). As a result, suppliers are more profitable than assemblers in Latin America—the reverse of the situation in the United States.* Even in the absence of controls, price surveillance by governments is much harsher on assemblers than on suppliers, with the assemblers given the task of holding components' prices down. The ability

---

* After-tax profits of supplier firms in the United States averaged the same as returns for all manufacturing—just over 13 per cent for the period 1947–65; the return for the three big manufacturers was over 25 per cent. A Brazilian study showed the reverse for 1966 and 1967; assembling companies reported returns of 7 per cent and 2.6 per cent for these years, while auto parts suppliers reported earnings of 12 per cent and 7 per cent. In 1968, only three of the auto assemblers earned a profit: 8 per cent for Scania-Vabis (Volvo trucks), 10 per cent for GM, and 13 per cent for VW. Ford Willys (a Ford subsidiary), FNM, Toyota, and Chrysler incurred losses. In comparison, only four of the sixteen parts firms had an earnings rate less than 10 per cent, and three had earnings higher than 25 per cent. Only one-third were foreign-owned, and their earnings were in the middle range. Two supplier firms in Mexico reported earnings around 15 per cent for 1968, compared to assembler returns considerably lower. And North American Rockwell Corporation reported that it received 72 per cent of its total foreign profits from Latin American affiliates, although only 58 per cent of its foreign investment is located there.

of assemblers to hold prices of suppliers down is limited by the monopoly position of some suppliers and by the necessity to purchase locally under content requirements. Only if there are a number of potential suppliers or if, as in the case of one Argentine assembler, the assembler has exceeded the local-content requirements can any pressure be brought to bear through shutting off orders.

Supplier companies differ in their attitudes toward regional integration. The assemblers that produce components are in favor for they feel they can integrate their component production. The large suppliers, who are foreign owned or who have a monopoly, tend not to be opposed but do not necessarily favor integration; they could gain from exporting to the expanded market, but they may lose monopoly positions. The small suppliers tend to be opposed for fear that competitive pressures will reduce their returns or drive them out of business.

The governments involved are generally not in favor of integration from the standpoint of supplier companies. The influences upon government officials are strongest from the larger, locally owned suppliers, where the economies of scale are likely to be the greatest and the pressures to concentrate production more intense; since these companies are not strongly in favor, there is no internal pressure. The eagerness of the assembler companies indicates to host governments that their control over the integrated companies is likely to diminish, increasing their reluctance. (Conversely, the fact that these companies have little local ownership means their influence with governments is small; as one Chilean official said, "The government is not interested in what foreign car companies think is good for Chile.")

To this reluctance is added the pressure of the small suppliers against integration. These companies have no technological advantage and various cost disadvantages—especially in high-cost labor compared to others in the region. To them, the opportunities of integration appear more those of death than life.

For the supplier companies to be made more competitive and have a chance to survive, they would probably have to become more responsive and closely tied to the assembling companies—accepting substantial assistance in techniques, scheduling, supply procedures, and quality control. Without some competitive pressure, this is unlikely to happen; with such pressure, the assemblers gain more control over the industry—a result not desired by host governments.

There are some steps toward integration that could be made without involving large numbers of suppliers, which at the same time achieve greater levels of efficiency. For example, integration in engine production or stampings would not involve existing suppliers extensively. Engine parts, such as pistons and rings, are made by suppliers, but these companies could be brought into the negotiations and given assurances as to shifts in supply patterns. The greater certainty that they would receive might be sufficient to overcome their reluctance to face some intraregional competition, especially in view of the fact that national economic growth would give rise to local competitors anyway.

Governments have not yet made up their minds as to how to handle local suppliers under integration. Public pronouncements are repeated concerning the necessity to integrate, and some countries, such as Argentina, have relaxed import restrictions in order to force cuts in domestic costs and prices. Any moves to integration in the automotive industry will require a decision on the place of local suppliers: protect their interests, meet their concerns part way, or disregard them. The succeeding chapter discusses ways of meeting their interests and still moving toward integration, by balancing benefits to foreigners versus nationals.

State Enterprises

The same trade-off between foreign and domestic interests arises in the petrochemical industry, but this time between governments and the international companies, since the suppliers of basic chemicals and some intermediates are frequently state owned. This fact poses a serious obstacle to integration, for the locus of decisions as to efficiency and equity solutions is shifted from the private to the public sector. State ownership and control of petroleum and natural gas resources has arisen throughout Latin America, for the purpose of guiding their development, controlling their refining, and sharing in their marketing. State control over petrochemicals is an extension of this governmental role; but not all countries have moved in this direction to the same degree.* Mexico has gone further than have Argentina or Brazil; it reserves a dominant role for PEMEX in the petrochemical industry.

If the three major countries were integrated, conflicts would arise over which company would control the integrated processes. The model discussed in Chapter 3 indicated that Mexico would become the center of production of basic and intermediate chemicals. This would place a large amount of control in the hands of PEMEX, or the Mexican government, with roles of Brazil's PETROBRAS and Argentina's YPF downgraded. In the search for efficiency, a trade-off arises in the loss of control by one government *and* an increase of control in the hands of another. For governments seeking to maintain or gain control over their economic growth, this would be a serious loss—one that might not be offset by the efficiency gains.

---

* See Appendix B.

In Argentina and Peru, however, the international companies could undertake their own integration back to basic chemicals. They would try to do so, in the absence of an intergovernmental agreement to the contrary, for their profit picture would improve by not being under PEMEX. But Mexico undoubtedly would refuse to join a scheme that permitted such a development to the detriment of PEMEX's role. Therefore, integration among the big three probably must define the role of the international companies such as to permit the governmental enterprises to continue to play their roles. Governments are unlikely to give way to the international companies or to another's state enterprise.

The same problem is made more complex under full LAFTA integration, for the center of basic and intermediate production shifts to Colombia.* Its state enterprise, ECOPETROL, is not equal in scope or power with that of Mexico. It is much more likely to rely on the international companies; and the resource base shifts to Venezuela, where the international companies are also dominant in petroleum and natural gas production. Even if such a shift were permitted, which is highly unlikely, the problem remains of what to do with the resources in Mexico, Brazil, and Argentina presently under development by state enterprises. And the government owns 100 per cent of the petrochemical industries in Chile, except for one operation owned 70 per cent by Dow. Would they be directed into uses other than petrochemical? Would this also require some type of agreement, facilitating such a shift in orientation?

In sum, the existence of state enterprises and the governmental policies that gave rise to them will require any integration scheme to provide for a meshing of the roles of the different state enterprises, a policy toward the ability of international companies to integrate vertically within the industry, or a means of redirecting activities of state enterprises away from petrochemicals if basic and intermediates are to be placed within a single national center. If centralization is not to occur, then division of basic and intermediate production must be provided for; this requires another trade-off against efficiency in order to spread benefits and control among nations.

International Companies

The problem posed by the existence of the international companies is a familiar one of can't do with them and can't do without them. As noted earlier, the Declaration of American Presidents states clearly that LAFTA integration is not acceptable if it becomes a means for the spread of foreign enterprise and domination of the economy by the international com-

---

* See Appendix B.

panies.* At the same time, these companies are presently the major source of technology and capital needed by Latin American countries for both automotive and petrochemical development. And they pose a threat to both auto suppliers and state-owned petrochemical companies.

Development of either industry to levels of international efficiency over the next decade will require reliance on foreign capital and technology and foreign technicians to train skilled labor. One attempt to design a purely indigenous automobile failed, and a couple of locally owned auto companies are still struggling to survive, but design costs are simply too high for so small a share of a small market. In petrochemicals, Latin America remains dependent on foreign technology at almost every level; even Argentina, which is the most advanced in terms of technical training, is reliant on such inflows.

Although Latin America could develop its own indigenous capital sources, it has not done so sufficiently to supply all the needs for industrial growth much less the special and additional requirements that would be imposed by regional integration in major industries. Successful moves toward sectoral integration would probably draw more international institutional funds. But even if the region were to develop its own or some new international sources of capital, the international companies would be quite reluctant to give up majority ownership of any foreign affiliate. This attitude stems from a reluctance to release advanced technology—especially the *latest* designs or processes—without control over their use and an adequate return. Neither control nor return are deemed adequate unless the international parent holds at least a majority, and some consider only 100 per cent ownership to be adequate. The parent asserts that (1) it is too risky to transfer secret processes over which they will no longer have control; (2) maintenance of quality control is necessary to keep their international reputation; and (3) the close meshing of the operations of affiliates with the parent and other affiliates, which increases efficiency, can be done only with majority or 100 per cent ownership.**

The magnet for capital flows, therefore, is the flow of technology. Capital will not flow without technological accompaniment, nor the most advanced technology without equity capital—at least not in the two industries under study. This relationship will obtain as long as there are opportunities that permit the tie-in; only when all other opportunities are closed off will capital and technology be separated in these two indus-

---

* "Mexican government officials argue that LAFTA would make little sense if non-LAFTA companies dominated its economy. No nation in Latin America, they contend, should serve as a launching pad for invasions for other LAFTA members by foreign-controlled firms." (D. M. Kiefer, "Mexico Strives for Industrial Independence," *Chemical and Engineering News,* Dec. 4, 1967, p. 99.)

** For further analysis of why this position is taken, see Jack N. Behrman, *U.S. International Business and Governments* (New York: McGraw-Hill, 1971), Chapter 5.

tries. Transfers of technology to Latin America also require heavy infusions of both skilled labor and management. For technology to be transferred readily, there must be a base of skilled technicians to receive it, adapt it, and utilize it for local needs. Such a base does not exist in adequate numbers in any country in Latin America, and to build one requires both time and foreign assistance. To be successful in shutting off alternatives and providing an adequate technical and economic base for growth, regional integration is necessary for all but possibly Brazil and Mexico.

Importing technology, capital, and personnel means that some of the benefits of industrial growth are paid out to foreigners. Integration, which relies on the international companies, will enhance the gains to foreigners. The trade-off facing governments is the increased benefit of more rapid industrialization and more efficient use of resources against the payments to foreigners; so long as the former are larger than the latter, the payments are worthwhile. But governmental policy is directed at maximizing the difference remaining within the country. The objective, then, is to gain the largest inflow with the least remittances, which usually leads to attempts to cut ownership while maintaining the inflow of technology and managerial and technical assistance.

There is really no economically effective alternative to use of the international companies open to the Latin American countries—whether they adopt a national policy or one of regional integration. The alternatives are to purchase technology and managerial assistance outside of the international companies or from them, *or* to develop their own technology and expertise, *or* to turn to Soviet assistance. The second route, of course, can be taken, but the time period is quite long—probably thirty to forty years, given the base of technology and training facilities in Latin America—and these countries would not be guaranteed the capability of keeping up even then. Japan has been successful within a twenty-five-year span largely because of intensive efforts and a substantial technology base. Even Russia has not been able to commit enough resources to technological development to match progress in the West in *both* military and nonmilitary pursuits; it would be a source of technical assistance but probably not an adequate one in terms of the available capital and the time required to accomplish a given project.

Purchasing technology means relying on licenses under patents, on expired patents, or on transfers under know-how licenses or technical assistance contracts. Although there are a number of licensing arrangements, including some in the automotive and petrochemical industries, the most advanced technologies are not usually transferred in this manner;* rather, they go to affiliates at least majority held. Patent licenses are relatively

---

* The most advanced techniques in petroleum refining are usually transferred by licensing, since they are built into the plant constructed for the customer licensee.

useless, for patents alone are not significant in the petrochemical and auto industries, and a license providing only the right to "manufacture, use, and sell" does not transfer enough technology to permit the profitable use of the license; know-how, training, sales techniques, and marketing and management skills are also needed. Attempts to rely on expired patents would merely mean that the investment would be in obsolete equipment and processes.

License agreements can and do provide for a continuing flow of new techniques and processes to the licensee, but they do not cover all of the processes used by the licensor, nor even all of the relevant ones (if these are the latest and most advanced being developed by the licensor). The licensees, therefore, will always be a step behind. Japan's maximum use of licensed technology from the United States and Europe helped build the base for domestic technological leaps sooner than it could have been built locally.

Management contracts that provide temporary managerial assistance and training can be a substitute for similar transfers under international companies, but the assistance may not be specific enough to the type of industry being developed. Generalized managerial services are useful in some contexts, but production know-how and customer service in technically advanced industries require knowledge that generalists in management do not have. Only if the recipient personnel know how to adapt the management techniques being received to the technical situation, can they build on what is being transferred. This also takes training and time. A trade-off is required between national control, maximum retention of gains, and the acceleration of industrial development. If the former two are more important, growth will be slower.

The fact that the international companies are the only enterprises already operating in more than one Latin American country makes integration both easier and more difficult. These now have the knowledge and expertise required to succeed in different markets and would be much more able to pull their activities together than would diverse national companies or state enterprises. The Latin American chemical companies are frequently not even in the *petro*chemical (as distinct from nonpetrochemical) sector and would have to shift the basis of operation first. Second, they would have to find ways of joining with other national enterprises—sometimes a state-owned company. If they joined with affiliates of the international companies, they would only strengthen the role of the foreigner; or, if they took over a foreign-owned affiliate, its contribution to regional growth would be weakened. Third, in petrochemicals, the strength of the international companies will be increased by the fact that they have concentrated in the intermediates rather than final products. Final products are economically produced close to a national market, while intermediates could readily be transported over Latin America. International companies, therefore, will respond more readily to integration than will local companies.

In the automotive industry, the assemblers would benefit most by integration. Some local enterprises might find partners in the supplier segments of other countries, but their monopolistic position would be eroded if strong suppliers remained independent in another country and could serve the regional market. Hence, integration is likely to strengthen the position of the international companies vis-à-vis local enterprises in both industries. Yet, despite the fact that the international companies would benefit from integration, they have not pushed hard for integration, for several reasons.

First, the international companies have not invested with a view to future integration; many of their operations are duplicative, being built behind tariff walls. Integration might require that they give up operations in one country in favor of expansion in another, raising the question of what should be done with the obsoleted facilities.

Second, not every major international petrochemical company in Latin America is in all major countries. It is not readily discernible how they would alter their operations (or expand) to meet LAFTA-wide integration. Contrarily, in the automotive industry, the major U.S. producers are in all major countries or could readily buy up smaller enterprises that do not have a regional complex. The automobile assemblers have unreservedly supported integration and would be able to respond quickly and effectively to a shift in production patterns, expanding trade among their own affiliates. Their slow movement in this direction is simply a response to indecision on the part of governments; no company wants to make heavy investments on the assumption that integration will proceed, only to find that they have merely created overcapacity.

A third factor is the potential role of state enterprises in the petrochemical sector. If, for example, the integration process in petrochemicals was left relatively free, the international companies would undoubtedly invest substantially in Brazil and Argentina, where they would be permitted to expand backwards toward basic production. This would undercut PEMEX's role considerably. Their next move probably would be to produce intermediates close to the basic products. Mexico would still attract international investment in final products, which are most economically placed near the markets. It is not likely, therefore, that the international companies would close their plants in Mexico; but their *growth* would occur in such a way as to locate facilities more in Brazil and Argentina where greater economics could be achieved by vertical integration. By this procedure, the international companies would probably be able to make their Latin American facilities competitive worldwide.

For Latin America, therefore, another trade-off exists; maintenance of state enterprises against international competitiveness and potential exports to extraregional markets. To opt for the latter means that the objectives of state control and local ownership (as espoused in Mexico) would be seriously undercut—and Mexican costs would rise. The state-owned PEMEX would not likely have the same capability to achieve efficiency by integrating across national boundaries within Latin America as do the international companies. Could it, for example, buy into or establish affiliates in Colombia so that the development of a center of petrochemical production there would be under PEMEX control?

Posed in this way, the question raises the more fundamental one of whether integration in the petrochemical industry should be under private or state auspices. That is, should the state-owned enterprises combine in some way to control the development, creating appropriate joint ventures? Or should they become adjuncts to a few international companies, which supply basic and intermediate chemicals and already have a nexus of affiliates over Latin America? Or, is there a third alternative in the creation of Latin American multinational enterprises—private or public? The division between public-versus-private enterprise has not been given a clear and decisive answer in any Latin American country for the economy as a whole and not yet for the basic industries, except in a few countries. The dividing line seems to be drawn in shifting sands. Yet integration requires that it be given a definite shape with certainty, especially in view of the substantial new investment that is required by the private sector, both domestic and foreign. The division between public-versus-private enterprise has only recently been made clear in a few countries (apart from the earlier example of Cuba). Chile has moved toward the public end and Brazil toward the private end of the spectrum, Peru seems to be following Chile, but the rest remain indecisive.

The actions of the international companies will affect this decision. The international company can get increased support from these countries by making substantial investments there: the host country gains industry that it would not likely have under integration and thereby builds up its bargaining power in future negotiations. However, the more the international companies insist on 100 per cent ownership in automobiles or petrochemicals, the more difficult it will be for governments to permit them to become the integrating force.

The bargaining position of the international companies is also strengthened by the fact that currently both Brazil and Argentina are providing them more leeway than Mexico provides, which feels that its attractiveness as a market puts it in a position to place some constraints on foreign investors. Under integration, the international companies could invest in Brazil and Argentina and still have the Mexican market. As a result of this competition, either Mexico would be obliged to give up its long-held policy of Mexicanization, or Brazil and Argentina would have to agree to become more restrictive.

The decision on the respective roles of the state, the international companies, and local private enterprises will substantially affect efficiency levels and also the distribution of facilities, trade patterns, governmental control over the industry, and policies toward foreign investment. For example, in the case of Brazil and Argentina, if the decision is to follow the Mexican route, relying on state enterprise, the economics of the petrochemical industry would favor Mexico as a base, shifting the growth of the industry there. Brazil and Argentina would gain more industry by adopting the private route, which lets the international company gain from vertical integration.

5

The Mechanism: Complementation Agreements

As shown in the previous chapter, the economic aspects of industrial integration that require resolution in a complementation agreement relate both to means of achieving efficiency and of channeling the benefits among members. Given the continuance of present insistence on national benefits and national control, the use of complementation agreements must not only promise acceptable levels of efficiency that include at least a measure of international competitiveness: it must also meet the diverse but specific national interests of each member country. The complex nature of the complementation arrangements and the difficulties of achieving required results often have not been fully recognized. If the problems seem too intractable or the responses too difficult, governments either will not integrate the more complex sectors or will take some other route to integration.

What follows here is not a prescription of what *should* be done but is a description of the problems that will be met under complementation agreements in advanced industries such as automobiles and petrochemicals and the responses that are available.

Economic Factors

The vital economic factors to be considered in attempting a complementation agreement—those relating to the problems of efficiency and benefits— include *location of production* (including the choice of specific products made in national facilities), *balance of payments, pricing, spread of technology,* and *employment.* Each of these decisions is left to market determination if duties are removed and no other arrangements made; or the market may be controlled by the international companies, which make the decisions. If these two means are not acceptable, another mechanism is required to gain an agreed-upon decision.

In coming to these decisions, the first move is to decide who will make the decisions—industry representatives or governmental officials. In almost all of the complementation agreements to date, the suggested structure has been presented to governments by industry (or company) representatives. The governments have, in effect, asked industry representatives—usually officials of trade associations or leading companies—to undertake the difficult task of negotiating differences. Governments then assess the results

69

and either accept, refuse, or request modifications. Even this first step requires prior knowledge of which countries are to be involved in the negotiations. Although all complementation agreements have to be open to entry by other members of LAFTA for a period of ninety days, it is difficult to make the bargains involved in balancing benefits if the "region" is to be redefined later. Similarly, if a country is permitted to opt out after having negotiated or signed an agreement, it upsets the delicate balance of benefits and may make the entire arrangement uneconomic, particularly if the country happened to be a singularly important market.

Means must be found, therefore, of inducing as many members as possible to join in the negotiations and of guaranteeing their continued participation. Inability to withdraw will make each member more concerned over the balancing of benefits to make certain that it will remain satisfied. Implementation of the agreement is equally important, to make certain that the balance of benefits agreed upon is maintained.

A difficult aspect of selection of members will be the inclusion of the lesser-developed countries (Paraguay, Ecuador, and Bolivia). All LAFTA complementation agreements are supposed to be open to their accession without concessions, even those that involve duty reductions on a conditional most-favored-nation basis. But an agreement that allocates production and dictates trade patterns must make concrete decisions as to the industrial roles to be played by these three countries—now and for a substantial period of time in the future. They have little bargaining power to gain a significant role by threatening to stay out, for that would mean perpetual stagnation industrially. Yet, to give them a significant role is dangerous to others, for there is no certainty that they could fulfill their projected part. Their inability to perform up to expectations would seriously alter the projected results of the economic factors discussed in this chapter.

Location of Production

Any complementation agreement that meets the problems discussed in the previous chapters must provide for allocation of specific production responsibilities among the members, perhaps even among companies. This allocation would be as precise or as loose as necessary to achieve the balance of benefits desired by (or acceptable to) the members. In some instances, the balancing may require quite specific allocation of components and subcomponents; in others, the allocation may be only at the level of systems or subsystems, with subcontracting by the individual companies permitted according to their own decisions.

In the case of an industry such as automobiles or petrochemicals—where the stages of production are closely linked and the potential number of companies is small—fairly tight interlocking of production must occur across national boundaries. Each unit must be able to expect fulfillment on schedule of the others' roles in order to be able to fulfill its own commitments.

Assuming that a decision were reached as to where production should take place, there remains the necessity to create the facilities called for. Some facilities might already exist, but in each of the countries—especially the lesser-developed members—new facilities will have to be created. Will private investors come forward? They may see the situation as technically or commercially different from that projected under the complementation agreement. For example, under the Andean chemical agreement (the sixth agreement), Bolivia was given the production of some rather sophisticated items, which seems to make little economic or commercial sense; Ecuador and Paraguay were also allocated a role.

It is conceivable that no investor will be forthcoming despite the provision that a given segment of the industry should be located in these countries. If, because of the difficulties of production, the operation is unlikely to be profitable, some inducements must be offered; these may or may not have been calculated in the cost of the production structure by the members negotiating the location. In the extreme, private enterprise may simply not respond and governments have to step in. Alternatively, some pressure to invest may be put on particular international companies, since they could arrange their affairs so as to maximize the benefits; this would strengthen the role of the international companies. The alternative is for the government to assume the role of the entrepreneur, taking the losses necessary to sustain the facility; or the other participants could agree to prices sufficient to cover costs of production in these less-developed countries (pricing is discussed in a subsequent section).

A second major problem with reference to location is how to restructure existing industry so as to meet the repartition requirements. Some observers argue that the only industries which lend themselves to complementation agreements with repartition clauses are those that are *not* already developed within the region. An industry with existing facilities in major countries poses too many obstacles, in their view. Nevertheless, a restructuring has been taking place within Europe based on established industries. The question is really how strongly government support will back a fixed structure of industry rather than a "fair share" in the industrial development. A sound existing industry provides the government with a strong bargaining position; but if the negotiators are really seeking an accommodation with each other, they will be willing to give up some of the existing strength in exchange for an over-all goal.

There is a cost to restructuring, however, related to disinvestment and investment in similar facilities elsewhere. In the automobile industry, it would be easy enough to restructure the assembly facilities, even if the present facilities were not located where they were wanted under the agreement. Some equipment would have to be moved, but that is not an exorbitant cost. A greater cost would be sustained in the movement of supplier facilities. Some would require consolidation to achieve economies of scale; others not. The most difficult decision here would be determining which country got the virtual monopoly position in stampings. Since only Brazil presently manufactures stampings, it would seem the logical contender. But to balance production facilities among claimants, Brazil would have to give up substantial other production now located there, and facilities would have to be moved. In engines, consolidation among companies would be required, though geographic relocation of facilities would not necessarily be forced.

In petrochemicals, much of the existing facilities would have to be moved if Colombia became the center for LAFTA-wide integration. If Mexico could hold onto basic production, fewer facilities would have to be moved; its extensive present industry would constitute a strong argument for leaving it there rather than pay the high costs of relocation. But, this would mean a balancing by shifting a large portion of intermediates and final products to other countries. The costs could be reduced by phasing out over some years; but the longer the period, the longer will be the delay in implementing the plan and the less efficient will be production in the early stages. Because of the high costs of relocation and the difficulty of sharing them, some observers feel it is too late to integrate automobiles or petrochemicals effectively.

Finally, since the location of industry at other than the most efficient places (as suggested in the previous chapters on petrochemicals) involves additional costs, the problem of the calculation of the cost of the trade-off becomes critical. It is possible that one repartition is less costly than another; if the two would be equally satisfactory to members, the least-cost one should be chosen. But, if the two are not equally satisfactory because of imbalances in the other criteria, which are discussed below, there are no guidelines to show how much should be permitted in lost efficiency in order to obtain equity in any other criterion. Recent economic theory has simply avoided careful analysis of "second best" alternatives because they frequently involve "nonquantifiable" objectives. A new analysis is needed to provide appropriate guidelines. Without them, the pattern of decision would follow that of the theory of games rather than that of a careful balancing of costs and benefits.

Balance of Payments

The impact of any complementation agreement on balance of payments is a prime consideration in acceptance by negotiating countries. The objective of each is usually to achieve a nil effect. The criterion of nil effect is usually a balance in imports and exports resulting from the new structure of production and trade. Exchange flows in the capital account or in invisible trade are not likely to be included in the measurement, though they are equally a result of the agreement; it is simply too hard to calculate them. A substitute for balancing of capital flows would be a further agreement on treatment of foreign investment (inflows) and intermember investment in facilities to be created under the complementation agreement. The effect of these capital flows will be to unbalance payments accounts. But balance in the trade may still be a satisfactory measure of equity among members.

A nil balance may not have to be the criterion, however, for all members have a payments deficit on trade account in the two industries. A country with a substantial import balance in automobiles or petrochemicals could not readily argue that the new arrangement should wipe out that balance; nor could one that had held its balance to zero through prohibitions on imports insist on improving that position. The opening of regional trade would undoubtedly result in an increase in import surpluses in some countries because of higher prices of imports from within the region (but lower than U.S. prices plus duty); an increase in volume of imports because of final price reductions (*ex* former duties) would also raise import costs.

The basic criterion that might be acceptable is that no country would be made worse off than before.* The complementation arrangement can provide the mechanism to achieve this criterion. However, it is doubtful that this criterion will be sufficient. It will probably have to be adjusted by a willingness on the part of some members to see investment funds diverted to the countries with larger initial payments pressures. The objective would be to assure the deficit countries that they would not remain in deficit as a result of the agreement. The new investments could well be made in items to be developed within the industry or which will shift as a result of expansion in demand and repartition of production.

Achievement of balance in trade payments within an industry requires exports to equal imports over some acceptable period. To achieve this end requires specification of volumes, prices, and schedules of delivery. Where a single, one-shot project such as a bridge or tunnel is involved, a precise balancing is feasible. But when each company responds to the market, such

---

* Welfare economics employs the same criterion, but without any explanation of how it will make good on this promise.

balancing is much more difficult. The efforts to balance trade under the Chilean-Argentine arrangement in automobiles have not been wholly successful. As noted earlier, an important reason was that the scale of production in Chile was only one-tenth that in Argentina but because of the uncertainty about the continuation of the arrangements, Chilean producers delayed making the investment in plant expansion that was required to balance the trade.

When such arrangements are expanded to include several countries and all components, the problems of meshing are compounded. A complex problem of input-output determination arises, further complicating the basic difficulties involved in restructuring industry to make it fit the new matrix. The matrix, in addition, could be extended to take into account the import content of the items allocated to each country; for example, stampings in Argentina and Brazil are made mostly from imported steel, but the stampings themselves are counted as 100 per cent local content. If a given component were allocated to a country and imported materials were a significant portion, the net payments effect would be significantly reduced. This problem would become even more important if the imported materials of subassemblies came from another LAFTA member but from outside of the auto industry and thus were not counted in the trade balancing.

A further element in the payments picture for each country is the transfer of earnings, royalties, and fees to parents of foreign-owned affiliates operating within the industry. Since the location of such affiliates within the LAFTA countries would not be spread evenly in proportion to production, some would bear larger payments burdens than others. Data for such flows are not available, but the magnitude could be significant: royalty payments by Mexico alone during 1968 were $6.5 million for automobile components, or just over 2 per cent of the value added by the Mexican motor vehicle industry.

Taking into account the direct imports of parts and components, the imports of materials not counted as "automotive" imports, the exports of automotive components, and payments as earning and royalties to foreign parent companies, the net foreign-exchange cost of the auto industry to Latin America during 1968 amounted to nearly $1 billion. For Argentina and Brazil, the import content of locally produced components was several times the direct imports; it was the reverse for Mexico and Venezuela. Chilean direct and indirect imports were roughly equal in value that year.

The main countries in LAFTA are therefore in quite different circumstances as to payments impacts before integration. Should they be in the same *relative* position afterwards? Would it not be reasonable to expect that their different levels of industrial development would lead to quite different responses in freeing of trade within the region and to different payments effects? In automobiles, for example, it could be expected that the heavier, more capital-intensive operations would be undertaken in the industrially advanced countries, for they have the supporting facilities. In petrochemicals, the same result is likely in the basic chemicals, unless the infrastructure in Colombia were significantly altered—at whose cost? These centers would tend to export more after integration, with a consequent large drop in their direct imports, offset somewhat by an increase in imports of materials.

The complementation agreement is supposed to rectify this potential imbalance. But it must do so *beforehand* and without precise information as to the flow of imported materials, the flow of capital, and the flow of earnings and royalties abroad. It will likely, therefore, focus on the pattern of trade. This pattern must be determined simultaneously with the decisions as to repartition of production of end products, intermediates, components, and basic materials. And these cannot be decided without reference to the volume of trade and prices, or the other equity objectives, such as the spread of technology.

Pricing

The level and variation of prices of the products in the two industries will affect the value of the segment of industry allocated to each country, as well as its trade balance. Given an extensive and precise allocation of production, with sole suppliers created in some countries for the entire region, pricing techniques become critical in the process of sharing the benefits and in keeping the final products competitive in the world market. A price committee would have to be created to set and alter prices among participating companies, for the complementation agreement creates a situation of bilateral monopoly or multilateral oligopoly, with governmental blessing. If production, trade patterns, and prices are set by agreement, a new set of multicompany relationships is created. The structure would not be like a cartel, for it would be composed of members serving each other before selling to the public, and those few companies at the final product stage would be under close surveillance. In both sectors, government-owned companies are likely to be part of the complex.

The closest pattern similar to what is projected may be the coproduction projects under the North Atlantic Treaty Organization (NATO) in which a common weapons system was procured by governments from a group of cooperating companies; the producing group was made up of independent companies located in the different countries and operated according to agreed guidelines. The guidelines covered the location of production, transfer of technology, the flow of components and parts, assembly operations, pricing, and final testing. Prices were either set beforehand or determined by agreed-upon formulae or agreed procedures; some final renegotiation was required in several of the projects.*

It is, of course, easier to determine prices for a fixed system to be bought on a once-for-all basis than to set prices for a continuing flow of production. Nevertheless, as noted previously, price controls are imposed on final auto assemblers and on many chemical producers throughout Latin America; since the suppliers or intermediate producers are usually not controlled, there is profit squeeze at the final stages of production. An extension of price controls or negotiations throughout all stages would be required under an intergovernmental agreement. It would be untenable to leave the setting of the price of a key component or intermediate to a sole supplier or to its government. The task is feasible if the regional market were closed to imports, but it becomes exceedingly difficult if prices are to be set that are competitive internationally.

Another factor that makes pricing guidelines necessary is the potential establishment of multinational companies—either foreign-owned or Latin American. Such companies, operating in the same industry across national boundaries, could alter the value of production and value and volume of trade through intracompany pricing. They might alter prices merely to meet the criteria of payments balancing. If one affiliate was exporting more than it should, it could simply underinvoice the exports to the other affiliates, recouping the losses at that end. A "balance" in payments could be achieved without any shift in production or employment. Under the Chilean-Argentine automobile arrangement, for example, one company reported that engine sales to Chile were priced at variable costs only, while Chilean parts were exported at prices three times the Argentine level. In another instance, the Venezuelan subsidiary of one company was attempting to convince the government of the profitability of permitting it to *give* parts to the Chilean affiliate to permit expanded operations in both countries.

Different pricing standards could be based on international prices, on production costs, or on some weighted cost formula. Any one of these would require a mechanism to define the item under consideration and the cost to be applied to it. Taking the motor vehicle industry, if the final prices are set by worldwide oligopolistic competition, the question is not how to set the final auto price but how to set prices on *components* so as

---

* The multinational consortia under NATO successfully faced on an ad hoc basis all of the problems that the complementation agreements would have to solve on a continuing basis. See the forthcoming study for the Bureau of Intelligence and Research of the U.S. Department of State by Jack N. Behrman, "Multinational Production Consortia and International Industrial Integration."

to divide the benefits equitably and to induce production of an adequate volume of each type. If prices are set too low, investors may not be forthcoming; if prices are set too high on some components, they will have to be set too low on others, warping investment decisions or profit returns. The task is easier in automobiles than in petrochemicals, for the intermediate and basic chemicals have alternative uses to being processed into a fixed number of final products. If prices are set too low in one area, producers may shift to another, upsetting the balance achieved under the agreement.

As a means of getting some experience in pricing, it might be necessary to defer the attempt to become internationally competitive and simply to set final-product prices at levels that can induce sufficient production of all components and then to experiment with cutting back on prices to reach international levels. One means of cutting back on prices would be to permit the price committee to open the market to imports of some components as a means of determining appropriate prices.

If international prices are used, then it is merely a matter of whether the country that has been allocated the production *can* produce at these cost levels. If there is adequate study beforehand of the industry and its potentials in each country, it may be known that local companies can meet the foreign supply price.* But, again, we run into the problem of the role of the international companies, which is discussed later. International pricing would tend to favor the use of the international companies, especially in the assembly stages, for it would be through the interlocking of their affiliates that they could adjust to the fact that prices set were out of phase with costs on some items. Even if prices were set in such a way as to benefit one country over another, the international company could absorb these differentials, offsetting losses in one against profits in another. The final cost would still be competitive and the final price profitable.

Pricing according to actual production costs would remove discrimination among various countries or companies that had no common ownership, since all transactions would be made at arm's length. The problem then becomes one of determining production costs. As found in the NATO experience, it is too difficult to reconcile the different cost-accounting procedure among countries. A common costing procedure would make sense in one country but not in another. Considerable effort is required in simply determining what costs should be included and how much. The problem of joint costs would arise in several auto-supplier companies, such as forging and machining, where the same capital equipment is used in non-

---

* This was found under the Starfighter program in Europe during the 1960's; the problem was simply how to do it. The "how" included a substantial amount of technology transfer, which may be the solution in Latin America—particularly in the case of the two industries under study.

automotive components. Differences in stampings for the several auto companies would require differentials in costs—how determined? Of course, private companies routinely solve these problems; the question is how to solve them jointly across national boundaries and to the satisfaction of governments.

Pricing by guidelines has been done for specific projects under defense orders in many countries. However, the procedure is largely related to a specific, one-shot project, and there is usually substantial renegotiation. Seldom is a company wholly reliant on one such contract or negotiation, as would be the case for those companies under a complementation agreement that allocated production and trade. Renegotiation would produce a high degree of uncertainty over a long time period as to the total revenue and costs of a company, which could not alter its output in response to post hoc determinations. Unless a high degree of credibility could be achieved by the control authority, and the rules or guidelines could provide adequate incentives, major enterprises would probably not invest under the new arrangement. It should also be pointed out that international companies have not stayed out of complex arrangements when they have seen a chance of expanding their operations and making a satisfactory return on investment of capital and management, even when these arrangements have involved strict regulations concerning ownership, prices, and markets.

Spread of Technology

Complementation agreements that redirect industrial development must also allocate the technologies to be used; this is implicit in the repartition of production. There may be some choices as to technology in production of a single intermediate or component, and some governments may have an interest in that choice. What has to be determined is where a particular technology is to be used. Both governments and companies are concerned over whether they will get a chance to share in the use and even development of a given technology.

It does not appear likely now that Latin American countries can be catapulted into technological development, for they have not made significant contributions in any field and have been quite late even in adopting technologies used internationally for some years. The lag in Latin American petrochemical technology ranges from five to twenty years. Such a lag prevents the users from making any new contributions to the same technology. What must be done first is to reduce the lag. This means a larger and faster import of technology from abroad—largely if not wholly through the international companies.

Eventually, however, Latin American governments will try to pull scientists and technicians into their economy and build a base for indigenous technological growth. Therefore, the allocation of technology among the LAFTA members will become an issue for negotiation. If the allocations are made on the basis of existing capabilities, the three larger countries, together with Colombia and possibly Venezuela, will be favored; they are capable of insisting on advanced technologies being located within them. But to perpetuate this situation by allocating to these countries advanced techniques, thereby stimulating indigenous development, would leave the lesser-developed countries still farther behind technologically.

On the other hand, not to use the existing capabilities in the advanced countries is to delay further the time of international competitiveness; capital assets as well as trained manpower would not be used effectively. What would seem to be required is a means of making a transition from the existing situation to a more balanced pattern of technological growth. But such a balanced distribution is quite difficult to achieve within a single industry without a substantial loss of efficiency. Trade-offs of techniques among industries would reduce the cost penalty. Economies of scale could be reached more readily, external payments balanced more easily, and some advanced technologies located in each country—though not one in each industry. Whether or not Latin American governments will be willing to accept specialization in certain technologies and to rely on others within the region to use or generate another technology is yet to be seen; such specialization will become a matter of negotiation if complementation agreements are pushed as a means of industrial integration. Even the Council for Mutual Economic Assistance (COMECON) in Eastern Europe has come to the point where the members have had to seek specialization among as well as within industries to achieve economies; it has not been at all easy for them to make the allocations since problems of national interests have arisen, similar to those discussed here. To determine specialization by agreement is quite a different process than to let it be determined by impersonal market forces; it is even more difficult than the process the international companies must go through if they were to determine the pattern of national specialization. To embark on such an endeavor, the gains should be seen as well worth the cost.

Employment Levels

Given the fact that it is possible to shift total value of production through pricing and that production location may be decided on the basis of the spread of technology, any complementation agreement creates another trade-off—one of efficiency against the level of employment and types of employment among member countries. One of the major concerns of governments in Latin America is the continued increase in unemployment and underemployment. Industrial advance has as one of its objectives the employment of increasing numbers of urban citizens. Therefore, any complementation agreement will have to provide for expanded employment and an appropriate division of the employment opportunities among participants.

A trade-off also arises between types of workers employed. In the petrochemical industry, acquisition of a basic-chemicals facility will mean high technology and will require a few technically trained workers, but it will not mean much in numbers of workers employed. Of course, there are employment opportunities in related industries, surrounding the development "pole," but not all of these can be counted in advance or attributed wholly to the one facility created. Similarly, one type of component in the automobile industry will give rise to greater employment than another—and of a different type.

How can these different impacts be weighed in order to achieve a balance in the negotiations? The same problem arose in negotiating some of the NATO production consortia. It was found so intractable that this criterion of equity was dropped. But, it may be that in Latin America the pressure of unemployment will be felt more strongly and that therefore the matter cannot be dropped. To date, however, we have no criteria for determining the nonmonetary benefits of one type of employment compared to another; nevertheless, such comparisons would be necessary to make the trade-offs.

In sum, several simultaneous decisions must be made in order to achieve an equitable solution among negotiating countries under a complementation agreement of the type under discussion. The complexity of these determinations may be such that the effort will be abandoned; or they may be made rather roughly, in a barely acceptable fashion, under the pressure to make some positive moves toward industrial integration and a level of efficiency that provides international competitiveness. In addition, some means must be found for organizing cooperative production and trade under the agreement.

Organization Alternatives

As discussed in the previous chapter, the existing institutions include the international corporation in both industries, protected locally owned suppliers in autos and producers of final chemicals, and state enterprise in petrochemicals. Any of these could be made the basic organizing entity for a complementation agreement, giving that group the responsibility to achieve an efficient and equitable restructuring of each industry. Alternatively, a new form of company—a Latin American multinational enterprise—could be established to pull the existing companies together.

International Companies

The international companies concern is to achieve least-cost production over the world. Since they are already in control of assembly operations in the automotive industry, their use would make it easy to organize production and trade regionally. Also, the problems of equity could be more readily met through balancing the activities of affiliates controlled by a few parent companies. For example, if the auto industry were concentrated in five or six companies—say Ford, GM, Chrysler, VW, FIAT, and Toyota—it would be possible to have an industry competitive both within Latin America and, in the case of some producers, internationally as well. For all companies to be internationally competitive, they would have to integrate with affiliates of the same companies elsewhere in the world. VW's operations in Brazil are probably the most efficient in Latin America, but it will still have to double its output to reach the level of 400,000 units, at which it might reach competitiveness. The other companies have operations no larger than 50,000 units, which would require both considerable concentration of companies in each country and integration across national boundaries to make each company efficient.

Supplier companies owned by foreign parents would have to be consolidated to achieve international efficiency. Governments could require that these moves by foreign companies result in an equitable share of production, no significant exchange drain, and an equitable spread of technology and of employment. The balancing among equity criteria could be left to determination by the companies with the final assent of the governments. Pricing could be left to the forces of oligopolistic competition within the countries and from outside. Under this approach, a variety of local suppliers would probably be induced to consolidate across national boundaries or to tie in with the major assemblers, or both. The results of the Canadian-American automobile agreement are instructive here, for a number of independent suppliers objected to the agreement and have felt themselves harmed; but both governments have found the arrangement effective, as have the major producers and their captive suppliers. Some assistance would be needed to help the independent suppliers make the adjustments.

A major advantage of using the international companies is assurance of a continuing flow of technology from the parent and other affiliates, including not only management skills, but also production processes and designs. Until Latin America is able to generate its own technology, it must rely on the major companies, even if a unique Latin American model is produced. Since it cannot do without these companies (or some of them), it may be most effective to use them in generating equity solutions also.

Another significant contribution is the marketing organization that goes with the international companies—an organization that can be spread quickly over Latin America and extends into third countries elsewhere in the world. FIAT's connections in Eastern Europe and Russia could lead to substantial swaps of components or models, some of which might involve Latin America; VW's affiliates might serve Africa; the U.S. affiliates could be dovetailed with others over the world. Governments in Latin America might have to give such exports a push, but they cannot rely on this marketing network without satisfying the international companies as to the mutual benefit.

Reliance on the international companies would certainly make for greater efficiency and even for easier solution of the problems of equity, under guidelines laid down by governments. But it is unlikely that governments will rely on them. There is such a strong current running against the foreign investor in some countries that governments would fall before they could hand over such decisions or operations to foreign enterprise. The Andean countries—notably Chile and Peru—would certainly not be able to accept such a solution; it is conceivable that the three big countries could, but Mexico has strong reservations about letting go of control.

An analysis of the petrochemical industry leads to similar conclusions. The major international companies could be the vehicle for complementation; but, as in the case of the motor vehicle industry, they are not currently set up to do so. The international petrochemical companies have not structured their investment to date in such a way as to take the maximum benefit from potential integration. A substantial restructuring would be needed with additional investments. They would have to regroup, consolidate, and concentrate facilities and ownership in order to become productive and internationally competitive. They are unlikely to make these moves unless it is clear that the members of the agreement will not renege on their membership, thereby restricting the future market.

These companies could and would make the necessary moves given the proper stimuli and assurances. The stimuli would have to include integration of substantially all the Latin American region—or at least the three big countries—and pressure to become efficient through selected imports from outside the companies involved; that is, they would have to be prevented from merely tying together a group of weak, overprotected Latin American companies and from avoiding import competition by locking in affiliates in third countries through imports, unless they were approved by the integration authority. Approval of tie-ins would be based on cost comparisons among local and other foreign suppliers.

The major obstacle to effective use of the international petrochemical companies as the organizing base is their inability to integrate vertically back to basic and some intermediate chemicals, where the greatest cost savings occur from integration. To gain the greatest advantages from economies of scale, they must be allowed to push back at least to ethylene and propylene production. However, the obstacle of government ownership of these facilities could be removed through pricing the basic chemicals at international levels.

An alternative way of using the international companies—one that would permit a slower move toward integration and a gradual testing of the results—would be to let each international company propose its own inter-affiliate arrangement for the region. Farbenfabriken Bayer A.G., Farbwerke Hoechst A.G., Dow Chemical Company, Monsanto Chemical Company, E. I. du Pont de Nemours & Company, Union Carbide Corporation et al would be asked to lay out an integration scheme for their affiliates throughout Latin America. The automobile producers likewise. The governments would then balance them as they saw fit, renegotiating with the companies as to the final arrangement. In this way, the companies would not be asked to get together, competition could be retained, and substantial efforts at efficiency would be induced on the part of each company. It is still possible that the scale of operations of any one company would not be sufficient to achieve international levels of efficiency. Too much efficiency might be given up for an easier means of obtaining *some* integration.

Whatever the potential drawbacks, the technique provides a step-by-step means of moving toward integration. Companies might well come forward with proposals to integrate only a few production stages. For example, Union Carbide has been negotiating with the governments of Peru and Venezuela to establish a company complementation agreement to produce polyethylene in Venezuela and polystyrene in Peru, with exchanges of the two products. The same end could be achieved by placing the two products on the national or common lists, but either could be revoked or delayed. Rhône-Poulenc's two affiliates in Argentina and Brazil trade between themselves under national list concessions, but since the concessions are revocable, neither affiliate has invested additional funds to take maximum advantage of the exchange.

In addition, a difference in governmental views as to the advantages of company complementation agreements has prevented their being used; the same criteria have not been used by all to weigh the advantages. Venezuela may place a high priority on employment while Peru is more concerned with the payments impact, and their concerns shift with changes in the economic or political situation. Union Carbide's proposal to Venezuela and Peru was negotiated over a four-year period and was finally rejected when the Venezuelan government decided that it wanted to produce all major petrochemicals in a complex at Moron prior to any moves toward integration in the industry.

Given the existing roles of the international companies in automobiles and petrochemicals, there is no alternative to integrating the industries around this business institution *unless* the governments wish to nationalize or to "localize" them by forcing local ownership and control. If the governments do not force either solution, they may use the international companies to integrate the regional market under some pattern guided by the governments, consolidating and concentrating activities efficiently and equitably; or they may let the companies integrate their regional activities with their world-wide activities, causing Latin America to specialize.

Under the first alternative, all activities would be encompassed within Latin America, but the final result might not be internationally competitive, depending on the number of companies left operating and their economies of scale. Under the second, competitive efficiency would be achieved, and Latin America could retain some control by insisting on having a role in determination of the specialization chosen for it. Given these choices on use of the international companies, it is more likely that Latin America would opt for the first and hope for a future level of competitiveness that would permit exports to third countries.

Supplier Companies

The local suppliers of automobile components could hardly become the mechanism for regional integration. They are too small, too specialized, and too thin in capital and management, and too many are owned by foreign companies. Even the forcing of joint ventures would not establish companies in which local interests owned a majority share of vertically-integrated operations.

The role for these suppliers under a restructured industry lies in mergers across national boundaries or in being part of an international company or a Latin American multinational enterprise (public or private). Mergers among suppliers which are now wholly independent would be quite difficult—as is seen by the absence of such mergers in the more closely tied economies of the European Community. The problems of how to mesh operations and management seem almost insurmountable, though such a meshing is technically desirable. It would be conceivable, for example, that a producer of radiators and tubing might shift to radiators while a similar company in another country might shift to tubing. But governments would have to arrange the shift and help facilitate it by reducing some uncertainty. It might be unwise for one supplier to rely on another in a distant country, or for an assembler to rely on an independent supplier in another country, given the delays in scheduling and transport that are likely. Those suppliers that were affiliates of international components companies would be readily integrated throughout the region, playing the same role regionally that they now do on a national basis.

If the international companies are to be the major vehicle for integration, it would be natural for some of the assemblers to try to absorb some of the suppliers, pulling them together across national boundaries. The international company could assist in improving schedules, quality control, and services by its affiliated supplier just as is done now, but the exchange across national boundaries would be greatly facilitated.

If it were necessary, in order to achieve efficient scale of production, to combine suppliers across national boundaries, each national government would be faced with the dilemma of losing control over this segment of the industry either to a company in another Latin American country or to an international company. It would probably find neither choice to its liking. If a Latin American government opted for the international company, it would strengthen the foreign element; if it opted for a regional company, it would strengthen a competing nation. The strength of dislikes becomes more important in such choices sometimes than those of the likes.

State Enterprises

State enterprises provide a potential alternative to the international companies in petrochemicals as the hub around which the industry could be integrated. But how can such enterprises be melded to form the nucleus (or nuclei) of integration? Leaving the major state enterprises *un*integrated across national boundaries would mean that there would be several competing centers: PETROBRAS in Brazil, YPF in Argentina, PEMEX in Mexico, ECOPETROL in Colombia, IVP in Venezuela, and Empresa Nacional del Petróleo (ENAP) in Peru. The result would not be much better than at present, if any at all. But to attempt to integrate them would require decisions as to which basic chemical production was eliminated in favor of expansion elsewhere, downgrading the importance of national control over the state enterprise. The Venezuelan and Colombian state enterprises have agreed to swap feedstock for basic chemicals on a barter basis, but no further integration is contemplated. It is, of course, conceivable that a new industrial center could be jointly owned by all governments, but the major complex would have to exist somewhere—logically either in Mexico or Colombia. However, Mexico is so far removed from South America that it would have difficulty integrating its military forces with those of the continent in the event of conflict with an outside power, raising doubts about the national security aspects of having the center so far removed. In terms of transportation, Colombia is not much closer to Brazil and Argentina.*

Even if the national security argument could be swallowed, and governments could agree on integrating the state enterprise, completely new international institutions would have to be created, raising legal as well as political problems for the region. This institution would have to be composed of state enterprises, joint ventures, and the international companies (to assure the inflows of technology needed at the intermediate level). Since not all international companies have the same relationships now with state enterprises nor similar facilities in each major country, complex trade-offs would be required to restructure the industry and create a single authority. The problem is more complex and difficult than that which faced the European Coal and Steel Community, for international companies were not involved. Control by a regional authority of companies within the member countries is one thing; control of affiliates of international companies is quite another, especially since certain contributions are expected both from the parent (technology) and from sister affiliates (export markets). To create such a structure is not impossible—just exceedingly complex and subject to considerable political jockeying.

Latin American Multinational Enterprises

The final alternative is the formation of regional enterprises, appropriately combining local, foreign, and state-owned enterprises. In automobiles, new Latin American companies would be formed by foreign and local interests that formerly held the assembler and supplier companies. This pattern would not add much in terms of economies of scale or efficiencies compared to organizing under international companies, but it would alter the source of control. Local interests could obtain the majority voting rights, either by greater capital contribution or by government fiat. Even if control rested in the hands of Latin Americans from several countries, national control would be lost.

The new structure might take the form of four or five new companies combining Latin American and international interest—Sud America-Ford, LAFTA-FIAT, or Toyota-Andina, for example. Several of the existing auto lines might be combined with local interests and retain their present models if they can be produced efficiently on a small scale out of relatively standardized parts. For a variety of reasons, this approach would concentrate on the regional (not international) market. One of the costs of this ap-

---

* For a further discussion of military and defense considerations, see Chapter 6, "The Will to Action."

proach is loss of the opportunity to integrate the regional system with the worldwide marketing facilities of the international companies. It is highly unlikely that these companies would open their facilities to joint ventures of which they owned only a minority.

In the petrochemical industry, a more complex arrangement would be required with local private, foreign, and governmental ownership combined into one company, cutting across national lines. Several such combinations could be created, centered on several state enterprises and one or more international companies with different local enterprises. Any one government could be associated with one regional enterprise or with more than one. Efficiency would be increased over alternative structures through a tighter vertical integration from basic chemicals to final products. But there is also likely to be overcapacity due to duplication under such an arrangement since it is unlikely that any one combine would want to give up a particular product line to another. Competition would arise between government-owned companies—possibly between companies owned by the same governments. New rules of competition and provisions for equitable solutions would have to be developed, as well as new legal forms.

Legal Aspects

To accommodate the process of integration described above several changes would have to be made in national regulations or statutes, including changes in the areas of trade barriers, taxation, incorporation, and rules of competition.

Trade Barriers

One objective of the complementation agreement is to remove all tariffs impeding trade in the industry concerned. But the rules of LAFTA do not include removal of nontariff barriers. These barriers encompass a variety of administrative regulations, subsidies, governmental purchase techniques, and technical requirements. For the automobile industry, requirements on local content would have to be harmonized. In the petrochemical industry, state enterprise is itself the ultimate trade barrier since it is often used to reserve the local market entirely.

Administrative regulations include customs valuation and clearance procedures, which can be so used as to make trade costly through delays and uncertainty. If the repartition of production is acceptable to all members, they would have little reason to use these delaying tactics, but customs of officials do not always respond effectively to exceptions in the general pattern. Surveillance will be required to make certain that delays and added obstacles are not imposed at the borders.

The use of subsidies would likely come under the complementation agreement to make certain that they were eliminated or used only in support of the objectives of the agreement. As indicated earlier, it may be necessary for the government to induce investment in a given country in order to fulfill the allocation of production; if so, it will probably use subsidies in some form. To the extent that competition exists among the companies within the industry, governmental subsidy could upset the delicate balance of benefits projected under the agreement by making one company (or its operations in one country) more competitive.

Government procurement policies would have to be coordinated. In the petrochemical and automotive fields, where governments are large customers, there would undoubtedly be a strong temptation to channel purchases to local producers. To the extent that companies were competitive, there would be a chance to favor a national producer. The military are unlikely to order trucks and vehicles from another national assembler; but suppose that these vehicles are not produced in the given country while another type is. Will there be pressure to alter the locally produced vehicle sufficient to serve the same purpose? Or will bilateral exchanges arise in which two governments agree to purchase their requirements from the other in order to assure a volume of local business equal to the military requirements? To protect the agreement as to balance of trade, such bilateral exchanges might have to be prohibited; this also calls for surveillance by the integration authority.

Government purchases in both petrochemicals and automobiles could be directed to one or another country through the determination of product requirements. For example, suppose that the military had a requirement for a type of rubber or plastic safety device: a given design could be put into production in one country rather than another. Will either put pressure on the designers? Should a common design for each type of equipment be developed so that it could be obtained more economically through economies of scale? Should there be a repartition of military items among the various national industries within the region?

Finally, the content requirements which presently vary from 40 to 98 per cent would probably be substituted by a regional-content requirement of 100 per cent, exceptions being made as necessary for a few specific items that might not be producible for a time.

Taxation

If integration is to occur through merger and consolidation of companies across national boundaries, a series of bilateral or multilateral treaties will be required on taxation of transfers of products, funds, earnings, fees, equity shares, sale and purchase of companies, and so forth. Avoidance of double taxation, as under existing treaties covering income and estate taxes, will have to be agreed upon.

In addition, internal taxation of products coming in as components can alter the flow of final products and the benefits expected by each country. Even taxation of the value added within a country at a rate different from that in other countries on the same or related products will alter the revenue benefits among countries. One country could increase its benefit at the cost of another. Therefore, some harmonization of internal taxes imposed on the production and distribution processes would probably be sought. In the NATO consortia, unsuccessful efforts were made to eliminate all taxation, since the systems were being bought by governments and each wished to prevent a revenue windfall going to another.

Incorporation and Capital Stock

The formation of any type of multinational enterprise—based on private ownership by Latin Americans, on state-owned enterprises, or on the international companies—would require new laws covering incorporation and issuance of capital stock. At the present time, any company must be incorporated under some country's laws. Would the newly formed multinational enterprises be under a LAFTA authority? Or would the national laws be so harmonized that it made little difference where the incorporation took place? In a sense, Europe has opted for both of these procedures, but there is as yet no experience under either approach, and it is not clear that a regional incorporation law would stimulate mergers on the part of purely national companies. The international companies would probably make use of such a law with the probably unwelcome result of increasing their strength relative to that of the national companies.

The issuance of capital stock is presently authorized only within each nation, under its own laws. In order to sell the stock of regional companies throughout Latin America, new laws and regulations would have to be adopted—even if the regulations were merely to remove present controls and limitations. But, more likely, governments would want the same procedures of reporting and issuance. Without such common regulations, the regional enterprises would tend to be owned by interests in the countries with the most efficient capital markets—Mexico, Brazil, and Argentina. This result would be unacceptable to the others.

If it were determined that a given proportion of national ownership were desirable in the formation of Latin American multinational enterprises, how would it be possible to keep those proportions among the members while at the same time permit ready buying and selling of the shares? What would prevent Mexicans from buying shares in a given company on the several markets and eventually owning the company? Suppose there were four major automobile companies, each selling shares throughout the region: could these not be bought by the same interests in one or several countries? The balance of ownership could be upset, and monopoly groups created. The integration authority would need means of surveillance and control.

Rules of Competition

The formation of monopolistic or oligopolistic enterprises in either of the industries would require governments to impose some rules affecting their competitive behavior. We have already discussed the problem of price controls, and it is essential that surveillance be established not only over the final product but also over intermediate exchanges, so as to prevent "gouging" by one company of others within the system. In addition, to gain as much of the efficiency benefits of competition as possible, control by one regional enterprise over another must be avoided as well as control over competing companies by the same financial interests in a given country. Equally, the same financial interests among several countries should be prevented from obtaining control over the major enterprises. Given the difficulty of surveillance and regulation, it is likely that governments would participate directly so as to thwart private monopoly elements.

Finally, some countries show an increasing concern regarding the proper use of advertising and the ability of foreign companies to create monopoly through offering better financial terms to buyers, the latter resulting from better access to financial resources. Since oligopoly would be the result of the complementation agreements, and since both advertising and financial power become critically important in such a market structure, participating governments would probably insist on guidelines as to proper behavior in these areas.

The necessity to maintain continuing surveillance and to promulgate a variety of guidelines or regulation points to the need for an intergovernmental administrative authority.

An Administrative Authority

To accomplish all of these changes—in the structure of the industries, trade patterns, ownership, and the legal requirements—governments cannot simply sign an agreement. A responsible authority will have to be formed—

much as the High Authority in the ECSC had to be created to oversee rationalization in those industries.

The major responsibilities of such an authority would be to maintain surveillance over the implementation of the various provisions of the complementation agreement, such as repartition, balance of trade, balance of technology, employment shifts and training, pricing, competition, tax remissions, and subsidies. To perform its duties, this authority would require a constant flow of information and close contact with the various companies in the industry and a capability of calling individual companies to task for violations. The authority should also have the power to call the member governments together to discuss emerging issues and would probably be given the power to issue regulations in specific areas, e.g., pricing and nontariff barriers.

The formation of two or more such authorities for each industry agreement would soon raise the possibilities of combining staffs and forming a central authority for all such agreements. Such a development would push the Latin American countries close to the formation of a common industrial policy—at least for key industries—similar to that proposed in 1970 by the Colonna Plan for the European Community. It is a new concept for Latin America, but one that holds too many advantages to be rejected summarily. The path is not yet clear, however, for Latin America still has to obtain several prerequisite conditions before it can hope to pursue integration successfully through sectoral complementation agreements.

The Prospects:
Prerequisites and a
Proposal

Even if the arrangements in the preceding chapter were developed adequately, complementation agreements could not succeed unless certain preconditions exist. The transport system must be efficient enough not to permit final costs to be competitive; exchange rates among participants must remain virtually stable; the United States and European governments must refrain from interfering in the restructuring of industry; and Latin American governments must honor their commitments. Absence of any one of these conditions could warp the pattern of integration so much as to make the final products noncompetitive in the world market or change the distribution of benefits unacceptably. The first section of this chapter assesses the significance of these prerequisites; the second assesses the prospects for achieving them as well as for creating the institutions needed for the success of integration through complementation agreements; and the third suggests an alternative approach.

Prerequisites

An efficient transport system could be built at the same time that integration proceeded; but noninterference from abroad, stable exchange rates, and governmental credibility will probably be required before the first steps will be taken.

Transport Costs

The patterns of integration projected for either the automobile or petrochemical industry, whether LAFTA-wide or subregional, involve substantial transport of components, basic and intermediate products, and final goods. Subregional integration of the three big countries alone would require ocean shipping between Mexico and the other two and use of either rail or ship between Brazil and Argentina. The cost of shipping basic chemicals is almost prohibitive, given their low value per weight, making it necessary to ship intermediates in order to allocate production over the region and to place production of final products near the consumer. Even intermediates are costly to ship, making transport cost a constraint on the pattern of integration. The present transport system for chemicals could add substantial cost to final products under any projected integration scheme.* But one chemical official estimated that adequate ocean transport could be provided with an investment program costing less than $50 million, $20 million of which would provide an adequate system for the west coast. This investment program, he avers, would solve the technical problems of transport by sea.

More of an obstacle are the political problems, for complex administrative procedures delay shipment inordinately. Although there is an intergovernmental committee attempting to facilitate movement of goods through customs and obtain more expeditious handling of goods in ports, little has been accomplished so far. The present regulations give governmental officials the responsibility for selecting the ships to be used in intracoastal trade. Governments often require numerous loadings and unloadings for inspections, which not only cause further delays but jeopardize the purity of chemicals. Since these are neither economic nor technical problems, they can be resolved when and if governments determine that integration of the industry is desirable. If the complementation agreement is deemed equitable, the movement of goods will be facilitated much more readily. However, an investment program for improvement of shipping and port facilities would still be needed.

In the car industry, final assembly also should be as close to the consumer as possible; substantial shipment of components and subassemblies among the participants will be needed to permit specialization and achieve economies of scale and the objectives of repartition. The pattern of production and transport is not likely to be all in one direction for each assembly line. The major assembly centers will be fed by suppliers or subassemblers that in turn are buying from other countries in the region; crossshipment and retransport will be required in some items, such as pistons for engines or parts for differentials, in order to achieve equitable distribution of production. The impacts of such a transport pattern will be different for each assembly line, depending on the costs (and time delays) in getting components to each line. The means of rationalizing the most equitable distribution of transport costs and the most efficient location of production are not at hand; adequate data do not now exist.

Some idea of the problem can be gained by looking at the costs of transport of some specific components. However, even these data could be altered by a change in the volume of shipments, which would increase greatly under integration. Ocean shipping charges are calculated on the

---

* The economic inadequacies and the political obstacles of transportation in Latin America in relation to integration have been assessed by Robert T. Brown, *Transportation and the Economic Integration of South America* (Washington, D.C.: Brookings Institution, 1966).

basis of revenue tons (i.e., 2,250 pounds or 40 cubic feet, whichever yields the larger freight charge). Charges for assembled vehicles are based on volume, but parts are charged by either weight or volume. Rates between Latin American ports for assembled vehicles amount to about 10 per cent of the value of an average automobile. But economies of scale would more than offset these costs, and if volumes of over 100,000 were reached, automobile companies would find it profitable to buy and operate their own ships, reducing the costs.

The cost of unassembled parts and components range from $18.00 per ton (Callao to Valparaíso or return) to $25.50 (Callao to Rio de Janeiro or return), with Rio de Janeiro to Valparaíso ($22.50) and other rates at intermediate levels. If a ton of parts was valued at between $1,200 and $1,500 (a full-sized vehicle weighs 4,000 pounds), the transport costs would amount to less than 2 per cent. Parts with a high ratio of volume relative to weight, such as tires and seating, would carry higher transport costs, making it less economic to ship these among many countries.

But the actual costs of carriage are not the largest transport costs imposed by a system of decentralized production. The more important ones are those caused by delays in obtaining supply parts, with inventories tied up waiting the complementary equipment. The delays may be caused by lack of port facilities, lack of ships, or mere variations in travel time between ports some distance apart. Given the requirement that only Latin American ships must be used in coastwide shipping, the frequency of schedules becomes a problem, and seasonal demand may require critical shipments to be put aside for a later ship. The frequency of service between Latin American ports by Latin American ships is poorer than between these ports and North American and European ports, as one would expect from the pattern of international trade. But the growth in regional trade has improved service between some ports. Robert T. Brown's study found service to be "relatively good" among the neighboring Atlantic countries—Argentina, Uruguay, Paraguay, and Brazil—and usually good among the Andean countries of Chile, Peru, Ecuador, and Colombia.* The links between the two groups are relatively poor, except for those directly between Argentina and Chile. Those between Venezuela and all other countries (except for Colombia) are also poor. Consequently, trade between the Andean and Atlantic countries often requires transshipment at some intermediate point, extending the time transit to more than three or four months.

If significant reshipment of components is required, the costs of meshing the items produced by members of integrated industry are increased. Overstocking inventories of parts which have to be imported is a costly way

---

* See footnote, p. 94.

of meeting the scheduling problem, and, it raises costs of some items significantly; but it may be less costly than the delays, for these cause an involuntary accumulation of nearly finished products or a breakdown in the production line. For example, one automobile company had 600 trucks on its lot without bumpers, which it eventually had to fly in from the United States.

The final cost related to transport is that of financing the goods which are in transit and those which must be kept in inventory—either to avoid delays in supply or awaiting the arrival of complementary items. Some companies have found that the financing costs are often equal those incurred in transport. One may estimate, therefore, that the total cost of transport involved in manufacture for an integrated industry would amount to over 5 per cent of the value of the vehicle; and, if the car or truck has to be shipped to the customer, an additional 10 per cent is incurred. It is unlikely that these levels of costs will seriously affect the decisions as to how to allocate production among countries within the automobile industry, but improvement of the facilities and service would be needed to make the final products competitive internationally.

Exchange-Rate Stability

Complementation agreements that determine the patterns of regional trade within the industry and directly or indirectly set the price of goods exchanged cannot readily operate under exchange-rate instability. Given the fact that an equitable distribution of benefits would have been determined through allocation of production, trade, and prices, it would be unacceptable to let changes in exchange rates alter this distribution. This constraint was readily recognized in the EC with reference to the fixed prices of agricultural products and was recognized (though not met) in the NATO coproduction arrangements. Control over the prices of either automobiles or final chemical products, which is largely the case now in Latin America, means that the costs of components and intermediates cannot be permitted to rise significantly without making production uneconomic. If one auto assembler, say in Argentina, finds that the peso is devalued, his competitive position is improved vis-à-vis others; this advantage would be offset somewhat by the rise in import prices of components. Conversely, if Mexico had to revalue its currency, automobile producers and suppliers located there would tend to sell less than before, upsetting the desired balancing of benefits. Or the cost to others would rise despite the fact that suppliers received no more pesos than previously.

These changes in exchange rates do not necessarily reflect the movements of costs and prices within the specific industry under the complementation agreements. If such changes did reflect precisely the changes in wages and prices within the given industry sector, one could conclude that no *real* change had taken place. But since this meshing is unlikely to occur, some real shifts do occur, changing production levels, employment, and trade. To correct or prevent instability in exchanges requires a close meshing of rates of inflation, and few Latin American governments have been able to control their own rate, much less mesh them with others. In the view of several Latin American economists, there can be no significant progress toward integration, especially by complementation agreements, until the inflations in major countries are curbed.

After exchange-rate changes, to rebalance the benefits as originally agreed upon would require a complex readjustment of the levels of production, trade patterns, and employment. Given the prior determination of these elements and the difficulty of making marginal shifts when the roles in each country may be quite different, the closest approximation to adjustment might be for the country benefiting to make appropriate payments to the others. In effect, the payments would represent the difference between the rate changes and the changes in wages and prices within the industry. But such calculations are difficult, and if they had to be made frequently—as would be the case in Latin America if the present pattern continued—a feeling is likely to arise that a balance of benefits could never be relied upon nor fulfilled.

Given the multiple controls existing over exchange rates presently, there would even be an argument initially over the "appropriate level" of current rates. An overvalued currency would tend to induce a greater allocation of items with export potential to that country, because of its unduly low level of exports in the past; an undervalued currency would have the opposite effect. But if the two were corrected at a later date, the benefits would be warped further still, with exports going to the country with the currency that had previously been overvalued. During 1960 through 1963, the seven major currencies in South America (excluding Mexico) fluctuated from overvaluation to undervaluation or vice versa. Sidney Dell estimates the swing from a maximum undervaluation of 41 per cent in Brazil to a maximum overvaluation in Chile of 38 per cent. Only Chile remained on the overvaluation side during the three years, having a minimum overvaluation of 11 per cent. The maximum swing from overvaluation to undervaluation was estimated at 66 percentage points in Brazil, and the smallest was 14 percentage points in Ecuador.* The existence of overvalued and undervalued currencies and the probability that they will be changed in the near future means that it will be extremely difficult (if not impossible) for governments to determine the balance-of-payments

---

* Sidney Dell, *A Latin American Common Market?* (New York: Oxford University Press, 1966), p. 165.

benefits of any given complementation agreement. Yet an almost precise balancing of payments and receipts is desired.

If governments could be persuaded to let trade-offs be made among several industrial sectors simultaneously, any country gaining from exchange-rate changes could be required to make faster reductions of duties in other fields or to give up some of its production and exports in a new sectoral complementation agreement. Such sacrifices would tend to take some of the advantage out of exchange-rate changes, and the more extensive the coverage of complementation agreements, the more pressure to keep exchanges stable—at least among Latin American partners. But stability might be based on a multiple exchange-rate system—either one for inside and one for outside, or multiple rates within the region for each sector, set according to the rate at the time agreement on each industry was achieved.

Foreign Government Interference

One of the espoused objectives of Latin American integration is to eliminate the "strangulation hold of the foreigner" over the economies of the region. The hold that foreign governments have through the international companies could be used either to frustrate moves to restructure a given industry, or, once integration was achieved, to frustrate given policies agreed upon within Latin America. The first interference could be through exercising the U.S. antitrust laws to prohibit the concentration of industry in a sector under or jointly with U.S. affiliates. The second could arise through any government attempting to dictate or prevent, through the parent company, certain acts by the Latin American affiliate.

Any of the methods discussed in the previous chapter that might be used to create a new institutional structure among producing companies in automobiles or petrochemicals could readily run afoul of U.S. antitrust laws. (They are much less likely to conflict with Common Market or European antitrust laws.) These laws would undoubtedly prevent any consolidation of one U.S. affiliate (a major U.S. automobile company) with an affiliate of another. In fact, companies are so wary of this possibility that they do not even discuss with each other the rationalization that all know is needed. (They are left with the alternative of designing "company complementation" arrangements which are likely to be unacceptable to governments.) Governments might force a concentration; but the U.S. government might oppose a joint venture while Latin American governments would be eager to keep in the United States the dual capital and technical sources that would come from a joint venture. They might even want a new company that encompassed several U.S. and European affiliates, with a majority locally owned. Such multipartner ventures could easily be considered in violation of U.S. antitrust law since they probably would create a monopoly situation.

In addition, one of the major world markets for the automobiles produced in Latin America, or even for components and accessories, is likely to be Eastern Europe and China. U.S. trade and technology controls still apply to these countries, and they are particularly tight over certain petrochemical technology used at the level of intermediate products. Situations may not arise under which the controls would have to be exercised, or they might not be exercised even if the regulations applied; but the controls still exist and could be used to stop a transaction by a joint venture employing U.S. technology or having a strong U.S. interest.

U.S. balance-of-payments constraints, though less severe with reference to developing countries, have been used to force repayment of dividends and to slow down an outflow of funds. Expansion plans within an integrated industry could be altered by the inability of foreign parents to contribute their aliquot share, either through retained earnings or new capital funds, because of U.S. government controls. Even though the parent company agreed to the expansion project, it might not be permitted to implement the agreement. Some European governments (notably England and France) have also imposed constraints on capital outflows.

Ready implementation of the complementation agreements would require that each of these constraints be removed from those sectors being integrated. There is no precedent for this being done, and it is not at all clear that the United States or European governments would exempt these sectors from statutory and administrative regulations presently applied. Without the exemptions, however, the objectives of the Latin American partners would be frustrated—either initially or after integration had taken place.

Government Credibility and Commitment

Two types of political instability threaten the success of Latin American complementation agreements: lack of credibility of official pronouncements and decisions, and unwillingness of a succeeding regime to honor commitments of a prior one. Comments on the subject by Latin Americans during a sequence of interviews were to the effect that if citizens within a country do not believe their government, how can they expect outsiders to? The unwillingness or inability of many governments to stick to their word or commitments has been evidenced, with respect to integration, in the repeated withdrawals of concessions—whenever it became evident that they would have some effect on trade patterns and domestic production potentials.

Further evidence exists, according to LAFTA officials, in the reluctance to bargain seriously over complementation agreements, despite repeated pronouncements of government officials in favor of such arrangements.

Private enterprises feel they cannot put much credence in the statements of governments because they may obtain an assurance or commitment from one official only to have another contradict it or refuse to follow through. For example, one government minister may approve a subsidy or rebate, but a subordinate or colleague may refuse to implement the approval. International companies entering a country have often been assured that given the small size of the domestic market, their profitability would be protected by official refusal to permit any other foreign company to enter —only to find within a few months after the start of operations that a foreign competitor has been granted the right to invest. The result has been that neither company is profitable.

International companies looking at the possibilities of investment after integration—such as under the Andean Pact—are given considerable pause before committing themselves to investments of the $10 million to $30 million required for a petrochemical operation, for example, because they do not feel that the allocation of production within the region will necessarily stick. Further, the assurances given by each country that they will import their requirements of given products from another are not believable in the eyes of some private officials.

The difficulty of accepting assurances from government officials is increased by the influence which certain local business interests have over government officials, causing a reversal or significant change in decisions. The largest or more powerful local interests (including foreign-owned companies, as in the automobile industry) may be able to influence a decision so as to prevent integration initiatives. Fear of this influence is enhanced by the feeling that even key ministers are corruptible within many Latin American countries.

Government officials have added to the uncertainty by refusing, so far, to specify the criteria by which they would judge whether any proposed complementation agreement would be acceptable. Though the criteria in Chapters 4 and 5 are widely offered, governments have not officially adopted them. Rather, they tend to insist that proponents of an agreement demonstrate "what's in it that's good for us," without giving any indication as to what they would like to see. This places a heavy burden of analysis of various trade-offs on the company proponents, who have no indication of the acceptability of these trade-offs. Companies tend to be reluctant to make such an investment of time and money for so uncertain a return.

The other facet of instability is the concern that "revolutionary" governments (as in Peru and Chile) will not feel obliged to fulfill commitments of prior governments. If some members of a complementary agreement opted out, the entire structure would be significantly altered, making investments less than profitable. Other members of a particular agreement might not be able to persuade the recalcitrant member to honor prior commitments; this inability would be more evident if the arrangement were among the smaller countries only. Many observers in Latin America have argued that Peru's adherence to the Andean Pact occurred simply as a means of getting support from the others to oppose the U.S. government in the International Petroleum Company case; without that need, Peru would not have agreed. Consequently, in the absence of continued strong pressure, support for the pact is expected to atrophy and implementation to be slow or nonexistent.

Prospects

The prospects for effective use of complementation agreements depend on the willingness and ability of the Latin American governments to develop a consistent and cooperative industrial policy to provide the prerequisites discussed above, and to create effective institutions to implement their agreements. Knowledge and ability exist. But the lack of a *will* to move strongly toward integration is evidenced in their industrial policies, in their attitudes toward the various institutional arrangements available, and in the differing attitudes toward industrial integration among the various interested parties, governmental and private.

Industrial Policies

Latin American governments have not adopted the most rudimentary of industrial policies as yet. The first step in forming such a policy is to define the role of the public and private sectors. India has taken such a step, as have Israel, Taiwan, Japan, the United Kingdom, the United States, Canada, and other advanced countries. In some countries, the public sector is defined leaving all else to private enterprise for development; in others, the reverse is the case. The second step is to establish priorities among industrial sectors as to the rate of their development. And the third is to provide some inducements to the private sector to follow the guidelines. France has taken the second step, as have Japan and the United Kingdom to some extent; and Japan has taken the third step. The Colonna Plan is an attempt to cause the European Common Market as a whole to take steps two and three, at least insofar as the advanced-technology industries are concerned.

Latin American countries have not yet taken a firm first step. As noted earlier, they have made no clear division between the public and private sector, save in a few industrial sectors. This fact dampens the willingness of private investors to commit substantial sums to fixed capital. Nor have these countries taken the second step of formulating policies for each sector. To date, industrial policies have been composed of largely unrelated ad hoc measures and not reflected a consistent over-all line of action. There have been few clear-cut objectives, and the requisite continuity has been missing. Finally, because they have lacked selective criteria, national policies have had an indiscriminate effect on the manufacturing sector as a whole, not shaping specific sectoral structures in a predetermined fashion.

Without a defined industrial policy, each Latin American country is wary of a regional sectoral agreement for fear that it does not know what it is trading off or what it might have in the future without regional integration. It prefers to hold its options open by not integrating. The existence of a firm national industrial policy would enable each country to know whether there was a likelihood that it would enter a given sector and in what way. It is quite possible, of course, that such a national policy might also be a deterrence to integration in those instances where several or more countries had the same national objectives with reference to a given sector.

What industrial policy does exist is highly nationalistic, aiming at the development of every possible industrial activity within the nation. Automobiles and petrochemicals are good examples of this approach, especially in the larger countries. Consequently, investments in each are being made almost wholly on the basis of nationally segregated markets. Each new plant that is erected is structured to that market plus some hoped-for exports—but not on the basis of imports of components from others.

Thus, the absence of movement toward integration is explained by the difficulty, if not impossibility, of determining the price to be paid for the benefits. A LAFTA official summed up the obstacles in the chemical sector as follows: (1) national governments have reserved the first and second stages of chemical production for themselves; (2) they have not yet decided on the price that they might be willing to pay in order to achieve the benefits of complementation agreements; and (3) the large number of products makes negotiations difficult.* The key reason is the second one. These costs of integration are in terms of national sovereignty; or, more specifically, in terms of national security, control over the economy, control over "our destiny," and protection of "our way of life." However, these four concepts are also too imprecise to permit ready determination of the costs involved in the loss of any of them.

* Interview; see also Gustavo Magarinos, *Evaluacion del Proceso de Integracion de la ALALC* (Montevideo: 1969).

In assessing their national security requirements, Latin American governments see potential enemies as existing both outside and inside the region, which makes the problem of trade-offs very difficult to calculate and decide. To meet the threat of a possible enemy outside the hemisphere, it would be desirable for the nations to band together in some mutual security arrangement, as under the Organization of American States (OAS), which binds North and South America. Of course, this would not help in case of a possible U.S. military action to protect U.S. private investment in the region, which some people profess to believe is a possibility.

The larger and more difficult problem is the assessment of the threat within the region. Jealousies among Latin American countries have not been eliminated, and the military in each finds a "foreign threat" useful in gaining new equipment, higher pay, and prerequisites. The power of the military in one country may well pose a seeming threat in the minds of neighbors who feel that the other's army might want to flex its muscles or that its government may want to create a foreign diversion to consolidate domestic control (six of the eleven LAFTA governments are under military control). There have been just enough flare-ups along borders historically to keep the fears alive. Colombia and Venezuela have warred over their border recently enough that these memories are still vivid; Ecuador and Peru have had conflicts. And Brazilians note that Argentina "always" carries out its army maneuvers along the Argentine-Brazilian border.

From the standpoint of industrial policy, it is to the advantage of the military to have a complete supply of all equipment within its own borders. The political power of the military, therefore, supports a national industrial policy. In both Argentina and Brazil, officials interviewed on the possibility of complementation agreements responded that the military of the other country would never permit it—meaning, in plain language, that their own military would not agree. In neither country would the military permit heavy trucks or vehicles to be produced wholly in the other country—nor engines and transmissions, though maybe bumpers and wheels.

Of itself, this attitude may be enough to prevent serious complementation agreements from being negotiated in either motor vehicles or petrochemicals. Agreements are feasible in home appliances, some machinery, photographic equipment, and so forth, but probably not in basic industry, important to the military.

Loss of control over the economy is more difficult to assess as a cost of integration. What is usually thought of is control over the pace of growth, over balance of payments, and over employment levels and location of economic activity. Earlier chapters have shown how these concerns can be met by specific provisions of complementation agreements. The national government retains control through the mechanism of the agreement, though the objective is regional growth as well.

Control over "our destiny" involves the capability of developing high-technology products within the national economy, providing diverse employment opportunities, providing roles for scientists and technicians, and opening avenues for creativity within the country for its own citizens. The complementation agreements in automobiles and petrochemicals are likely to centralize much of the technological advance within a few countries. The question is, would the distribution of technology really be significantly different without complementation? Could national efforts succeed in drawing such high-level technology to the smaller countries? Or will they be left in a technological backwater any way they turn, unless they specialize among the advanced industries? Potentially, the three major countries can obtain high-level technology in both petrochemicals and automobiles but the smaller ones would have great difficulty in acquiring it in either. Even the major countries may lag in obtaining the latest advances, as the more-advanced countries add new technologies.

Protection of "our way of life" is a much more nebulous concern, but nonetheless real. The transition from slowly growing or traditional societies to faster-growing, more-industrialized societies is disturbing: values, relationships, and cultural patterns come into conflict.*

Psychological barriers arise from the differing national backgrounds and cultures of the peoples in the various countries. The Mexicans consider themselves apart from South America, having had their revolution earlier and having showed that they can develop a stable "democracy" long before the others have. Being largely European, Argentinians consider the Brazilians as "Celts and Negroes." The Brazilians have also had a major awakening in the realization that the rest of the continent is not really willing to try to understand or use Portuguese, leading to feelings of rejection. The Chilenos, who are mostly European, consider themselves a cut or two above the Peruvians; and the Colombians know that they are the most cultured group in Latin America. During an effort in the early 1960's under the Alliance for Progress to get Latin American industrialists to help each other managerially and technically, each national group gave the almost universal response that "we'd be glad to give them some help, but don't send any of them to help us."

These conflicts are intensified by the pressure from Europe and North America, through the international companies as well as through the communications media. In Latin America, the tensions from such change can be turned against the "foreigner in our midst" who is, obviously *a* cause of the conflicts. Foreign investors and the international companies can be-

---

* For analysis of Latin America that is helpful in understanding the pressures of change on this "way of life," see Ted Geiger, *The Conflicted Relationship* (New York: McGraw-Hill, 1967), Chapters 6–7.

come the convenient scapegoat (a useful concept for any government, and it is most readily used in the context of Latin American politics).

These four reasons for national industrial policies also support a policy of reducing the impact and influence of the international companies. To move toward complementation agreements, however, makes these companies more influential. Such agreements are likely to be acceptable in the key industries only if means can be found to downgrade the importance and power of foreign enterprise. Development of a regional industrial policy, which would underlie complementation agreements, waits on resolution of the above four desires on the part of national governments and their handling of foreign-owned enterprises. The conflicts can be more easily resolved if appropriate institutions have been established.

Institutional Means

Various institutional arrangements can be employed to implement a complementation agreement. The choice depends greatly on what equity criteria are selected and how they are to be met, on the extent of control desired by governments, and on the level of competition envisaged for the industry. Having examined the implications of equity criteria, let us now turn to the questions of control and competition.

Given the need to make certain that the agreements are implemented as agreed upon, the creation of a regional authority, as discussed earlier, is to be anticipated. The critical issues in its formation would be the composition of national representatives, their voting rights, and the executive's powers of surveillance and of command. Since not all members will have the same economic interest in a given industry, it would be conceivable to provide for weighted voting. But by what formula? The existing value of production in each country or that projected under one scheme of integration or another? And who would exercise the votes? Would industry representatives be appointed, and, if so, would they have voting rights equal or subordinated to those of governmental officials? What of labor representatives? How would votes for industry and labor be apportioned among the stages of production (basic, intermediate, and final products)? Finally, what of the rights of consumers in such an organization?

The problems of representation and voting rights have been resolved in other industrial organizations, such as the ECSC and the coproduction projects under NATO; but in neither instance is there quite the complex membership that would exist under the complementation agreements. In Latin America, the range of country participants includes some with quite advanced industrial sectors and some without any segment of a given industry.

But to give the advanced countries undue power in the commission would undoubtedly lead to dissatisfaction about potential future benefits. It is likely that these issues of control, voting rights, and representation would extend the negotiations on complementation agreements over several years. But given the will, they can be resolved.

The problem of control is intimately tied to that of the competitive structure anticipated under sectoral industrial integration. In the automotive industry, for example, the competitive structure sought could be that of an oligopolistic market in final products, with the sellers being the assemblers of the vehicles. Behind them, some suppliers might be consolidated into sole suppliers (over which some control would have to be maintained by the authority) and others might be numerous enough and on sufficient scale of operations already to permit relatively open competition; still others might wish to concentrate their activities under fewer companies, with the authority holding the right of approval.

The resulting institutional structure, as it might appear, is shown in Figure 6-1. The industrial authority would have direct responsibility for implementation through the international companies and control over any independent suppliers that raised problems through their pricing, trade patterns, employment levels, or bankruptcy. Suppliers affiliated with the international companies could, of course, be controlled through these companies.

An alternative means of control and of establishing competition would be for the various governments to take control over certain segments of the automotive industry in their country, or to buy into them on a minority or majority basis, thus gaining greater information and control than previously. In this event, the regional authority would be dealing with governments as well as international companies, and the latter would be supplied by state enterprises. The pattern would begin to resemble that under the petrochemical industry.

Given the existence of state enterprises, the petrochemical industry could be organized competitively along one of three or more alternative patterns, with the industry authority at the head. One would rely equally on the state enterprises and the international companies to organize the industry; another would rely largely on the state-owned enterprises to organize the industry; and a third would add the local companies in at the same level (see Figure 6-2).

The more detailed secondary relationships in A could exist under B and C also. The first industry structure (A) most nearly continues the present situation, with the state enterprises supplying the international companies and competing with them as well, even preempting them in some products at the intermediate and final stages. The authority over affiliates indicates that both major company groups might acquire subsidiaries or associates

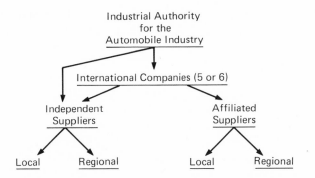

Figure 6-1. Structuring the Automobile Industry: *"Competitive Oligopoly."*

in other countries in the region that would need to be included. Mergers among the international companies or among state enterprises and even jointly held affiliates would be conceivable within this structure.

The second alternative (B) places the state enterprises in control of all elements of the industry in each country, with the government buying into the international companies on a minority or majority basis. The state enterprises might also buy into local companies. The ties among the state enterprises themselves become critical in this structure, including their ownership of different pieces of the international companies, fortuitously located within their jurisdiction. This alternative raises the question of whether a regional intergovernmental agency would not be required to hold the shares of the state-held companies so as to remove conflicting controls and to enforce integration. This, of course, would eliminate virtually all competition except possibly at the final product stages.

The third alternative (C) would attempt to create strong local enterprises, through merger nationally or regionally to stand up in competition with the other two entities. This movement would be guided by the authority so as to generate a sufficient level of competition and still meet the criteria of sharing required by the agreement.

Yet another alternative is feasible for either of the industries under discussion; this would involve restructuring ownership and control into new Latin American multinational enterprises, owned in partnership by governments, local investors, and international companies. Such an approach would fit into the pattern of B or C above. In each, the regional market would be the prime one to be served. The international market can best be served by giving the international companies precedence or at least full control over their own production; only if they can reap an adequate reward for the use of their distribution facilities could they be induced to open them to jointly owned ventures, and the reward is likely to be rela-

A. Equal Reliance on State and International Enterprises:

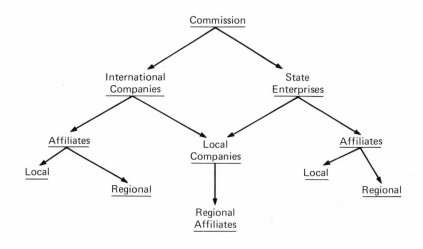

B. Reliance on State Enterprises:

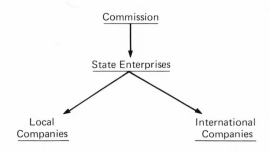

C. Equal Weight for Local Companies:

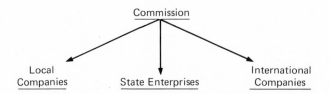

Figure 6-2. Structuring the Petrochemical Industry.

Figure 6-3. Model for Industrial Structure Serving the Regional Market.

tively small if their ownership is less than 35 per cent. Equally, adequate inducements will be necessary to cause new technology to be transferred.

However, if the Latin American countries are sufficiently intent on reducing the control of the international companies and are willing to focus predominantly on the regional market, the institutional structures of the two industries could be given still another form (see Figure 6.3). The multinational enterprises would include the state enterprises, producing the basic commodities, the intermediate producers (largely the international companies and some locals), and a few major final products producers (though this step is not necessary). Variations come to mind readily: the enterprises could be nationally based or might cut across the regional market combining companies of different nationality. The result could be a group of four or five major companies, each with their basic and intermediate production, competing throughout the region with each other and becoming efficient enough to sell in the world market, at least in some lines. This result would depend, of course, on the scale of operations and relative efficiency. Pressure could be kept on them to keep prices down through gradual elimination of duties against outside countries.

Ownership of the Latin American multinational enterprises could be broadened to include not only governmental and private interests in a given industry but also international financial institutions and aid agencies, with these latter phasing out once the operation is successful.

Will to Action

The complexities of the trade-off involved in a complementation agreement and the choices of the institutional means of implementation would seem to be so great that national governments might wish to seize on the first-best solution of maximizing efficiency and let the market make all determinations. But the promised maximization of welfare to the entire region is still not sufficiently enticing to make governments or national groups give up control over their economies, destiny, or way of life. Given the choice of national identity or wealth, nations will forego wealth. National control (identified with the concept of sovereignty) is itself a value, and it is not necessarily maximized along with the economic values under free-market industrial integration. The will to move along the route prescribed by the economists simply does not exist and is unlikely to arise.

What will there is to move toward integration is circumscribed by a strong desire to assure national benefits. What the previous analysis has shown is that, despite the complexities and the difficulties of making precise balances among various criteria, there are means of achieving a large share of the gains of integration and an equitable distribution of those gains. Trade-offs are necessary; they have been made in other programs involving industrial cooperation. The LAFTA situation is a good bit more complex, and a solution will require substantial institutional changes. But the task is feasible, though it might lead to a completely new form of industrial association.

The alternative routes are visible, and equally obvious is the fact that other industrial agreements have succeeded. What is lacking is the will to embark on the journey; this will is necessary before a route can be chosen. The lack of a strong will to integrate through complementation agreements is evidenced by the attitudes of the three major players: governments, local industry, and the international companies. An initiative must be taken by one or more of these, but none has seriously assumed responsibility for promotion of a complementation agreement in either automobiles or petrochemicals. Rather, their attitudes have been largely those of waiting for the other to take an initiative and then opposing what has come forth sporadically.

Governmental Attitudes. There is one complementation agreement in chemicals of the preferential type (the fifth agreement); there is a second one in petrochemicals under the Andean Pact, involving repartition (the sixth agreement).* But there is no complementation agreement in autos (nor have autos been included in the national lists even for small reductions in duties). According to LAFTA officials, the fifth agreement indicates the stance of governments: they will negotiate agreements only in sectors or for products where there is little or no current production or prospect of it. Even so, the fifth agreement relates only to a small list of future products and does not require allocation of that production among the members. Any agreement on products now in production would require allocation of facilities, but this task is made seemingly impossible. So long as there is any prospect in the near future of production within a country, that nation will not give up this opportunity under a complementation agreement.

---

* Described in Chapter 2, "Complementation Agreements."

Even within the four years since compilation of basic data on the petrochemical industry by LAFTA and ECLA, beginning in 1967, governments have moved rapidly to expand their country's participation in key industrial sectors. For example, in 1967 only two countries had polystyrene production; by 1969 all major countries were producing it. Each seems to want to be self-sufficient in raw materials, basic chemicals, and intermediates, so that they can move into any line of final products. To get a facility producing intermediates, Venezuela has decided to erect a polyethylene plant that can sell to the world market (and Latin America if permitted); it is not waiting for regional integration.* One international company offered to erect production facilities in two countries—Peru and Venezuela —exchanging the products on a balanced bilateral basis, but it could not get either of the governments to agree to rely wholly on the other supplier for the products; thus, the market remained too small to induce the investment.

Peru's new national industrial law has tended to create artificial industries, for which protection will have to be continued for some time; the orientation of the government has so frightened industry that little investment has taken place, let alone in regionally integrated facilities. Venezuela does not want to have to buy from others if quality is low or price is high, and any significant shift in trade patterns would offend some of its established suppliers (including Japan) who are also good customers.

Further evidence of governmental reluctance to integrate is seen in Brazil's granting to Uruguay an opportunity to export plasticizers, in which Uruguay could not possibly compete and in which Brazil had 200 per cent overcapacity. But Uruguay bought Argentine PVC at dumping prices and converted it into plastic pipe, selling to Brazil below Brazil's cost. Brazil then withdrew the concession, for it couldn't take the chance of the duty reduction actually leading to imports. In fact, a request for a concession frequently triggers an initiative by local industry to produce, for a market has been identified. Negotiations on duty reductions have become stuck on the problem of bilateral balancing of trade under concessions; any company wanting a concession must develop an opposite flow of trade so that governments will be satisfied. Some company officials see an eventual move to quotas in petrochemicals to force appropriate balancing of trade flows. But bilateral balance is not a sufficient condition to get governmental approval.

Despite a bilateral balance in a proposed exchange of automobile en-

* Union Carbide offered to erect the plant with only 100 million pounds capacity, even though its other plants are no smaller than 200 million pounds and one in the United States has a capacity of 1 billion barrels. The company calculated that with cheap natural gas it could keep costs low enough to sell some 75 million pounds to the world market. After this offer, Instituto Venezolano de Petroquinica (IVP) decided to do it alone.

gines between Argentina and Brazil, neither government would approve, for the exchange would have meant that one specialized in four-cylinder and the other in six-cylinder engines. Neither was willing for fear that the other country (and its military) would gain something not foreseen. Nor were they willing to let one U.S. company produce left-hand side panels for trucks in Argentina and right-hand side panels in Brazil and swap, though the Brazilian facility was producing one month's sales in two days of production. Both proposals were made for exchange between affiliates of U.S. companies.

Competition between Argentina and Brazil has been so intense that anything which one has the other wants: In autos, Brazil is three to five years ahead and Argentina is trying hard to catch up. Neither will agree to integration by car models, though this is feasible since both have the same content requirements. In addition, Argentina counts Brazilian components as Argentine-content if the same value of exports is sent to Brazil by the importing assembler—even if Brazil does not make the same concession. One assembler wanted to locate all production of a certain engine for passenger cars in Brazil and a truck diesel in Argentina, but both governments wanted the passenger engine.

Mexico is reluctant to integrate until its import content is equal to others; it is hoping others will come down. Mexico will not permit imports to count as local content, even if it exports an equal amount, for it wants to build up *net* exports. Besides, exports from Mexico build up production quotas for assemblers, so that exports for equal imports would serve to increase both content and quota—neither of which the government wants. This was one reason why a proposal to Mexico by Chile was killed. Chile had wanted Mexico to join the Andean group but later proposed a bilateral balancing of exports and imports of components on condition that Mexico deposit funds in Chilean banks sufficient to guarantee a balance of trade (which had not even occurred at that time with Argentina and never really did). Chile wanted a governmental agreement—not one between companies approved by governments. Mexican suppliers opposed the arrangements, for exports of engines would help the Mexican assemblers, but the imports would cut into supplier production since Chilean component production was identical to some of the Mexican production. The suppliers did agree to imports of two items that were not then produced in Mexico; after they were produced in sufficient volume, imports could account for only 30 per cent of sales. But the proposal was rejected because of supplier opposition and refusal of the government to change its regulations over exports and production quotas.

The general orientation to industrial policy has been to stress expansion of output through exports stimulated by government incentives. Brazil has given 30 per cent incentives to export of refrigerators; as a result, sales have been made to Ecuador. By 1980 it should have an internal market of 1 million vehicles, with VW producing nearly half—a quite efficient level. Some export incentives, such as removal of the 15 per cent internal tax and some small additional subsidy would make it competitive in third markets (it is already selling engine blocks to South Africa). This route is much more desired than getting tied up with other Latin American countries. Consequently, Brazilian officials are looking to make it on their own. A projection for the year 2000 shows a quite satisfactory situation for Brazil without integration; they are willing to wait. One Argentinian official stated in an interview that they could make it on their own in ten years and that "ten years is not a long time in the life of a nation." In each country, success still requires a reduction in number of auto assemblers and models made by each and substantial rationalization in the petrochemical industry.

The will of governments to move toward integration is affected also by broad economic and noneconomic factors. The broader economic ones relate to the pressure of overpopulation, unemployment, and income distribution. Any economic activity which merely raises efficiency without helping solve population pressures on the cities and mounting unemployment and poverty will have a low priority, for the span of government attention cannot stretch over many difficult problems.

Governmental decision making is complicated further by noneconomic pressures, political and military as well as psychological. The political will is affected by the desire for autonomy and a need to develop national unity before taking the step toward regionalism. Many of the countries are not well-integrated internally. Brazil is still divided between the coast and the interior and between the northeast and the south; Peru is divided between coastal inhabitants and the Indians of the Andes; Mexico is still striving to integrate the *campesinos*; Argentine politics are largely controlled by those in Buenos Aires; and Colombia is faced with rivalries among its five major cities. The political and military tensions among the countries, which have been discussed earlier, also add to the lack of will to integrate. If these attitudes are to be overcome, each participant must feel that there must be high rewards for integration and also that these rewards will be certain.

Local Companies. The reluctance of governments is paralleled by that of the local companies, particularly in the automotive industry. Though many industrialists have given strong verbal support to regional integration, they tend to send lower-level officials and technicians to Montevideo for LAFTA negotiations, people who are uninstructed as to the long-range objectives of their companies (which, of course, may not have any long-range plans). They could not argue, therefore, for a particular future pattern of production or trade, and they are often afraid of giving away an opportunity in a future product line. Consequently, the negotiations go on interminably over minute points.

In automobiles, the supplier companies see no advantage in opening up the local market to competition from other countries or of having to face the greater uncertainties of exporting to other Latin American markets. They prefer the stability (such as it is) of the domestic market with less competition and maybe lower profits (but higher profit rates). Industry in Venezuela, for example, has opposed joining the Andean Pact despite the fact that many companies are operating at 75 per cent of capacity; they would rather take the growth rate of the internal economy than face the uncertainty of competition in the Andean market. This fear experienced by supplier companies is based in part on a fear of the unknown—even if they might benefit, they simply do not want to face change. They also fear even stronger competitors in the form of affiliates of international companies that could be established in the lower-wage countries of the Andean group. On their part, Colombian businessmen want Venezuela in the Andean group, but only *after* Colombia has built up its competitive ability to meet Venezuela's highly capitalized facilities. Peruvian industry was also against joining the agreement.

Some American companies proposed a bilateral exchange of engines between Brazil and Argentina that would not seriously have affected the local suppliers; the latter would have had to shift sizes of some components but could have had longer production runs. Other industrialists opposed the arrangement. First, some of the engines would have carried carburetors with them, others not, affecting a few local suppliers. Second, other companies would have liked to have had a bilateral agreement but were not in the same position to exchange engines. And third, neither FIAT nor VW could take advantage of such an exchange, and they are the largest producers in Argentina and Brazil respectively. This very fact means that a proposal for a company complementation agreement is likely to be opposed by others in the industry (covertly, if not overtly) since it would give the initiator an advantage. Still, Toyota is reported to be trying to set up a four-country production scheme based on interlocking supplies.

Similiar problems arise in the petrochemical industry since major companies are not spread evenly over the major markets and existing local enterprises do not want to give way to imports. Brazil, Argentina, and Mexico have almost a complete line of chemicals, but there is much to be done in rationalizing the industry, filling in the intermediates, and making production competitive. Until these goals are accomplished, industrial representatives are not willing to talk about integration, which they support only as a means of opening export markets; to gain from such an opening without facing greater imports, they must have the domestic market locked in. This attitude has been strongly influenced by past experience with tariff reductions that have foreclosed production to local industry. For example, Argentina and some other countries were persuaded by Mexico to cut the duty on gum resin by 20 per cent. Mexico did not appear to be competitive even at that level of duty, and the concessions were granted. But Mexico then tooled up to produce at lower costs (some 2 to 3 per cent below the others) and absorbed the entire Argentine market, which local producers had hoped to enter. The prospect of future domestic production is used by industry to dissuade governmental officials from making concessions or negotiating a complementation agreement.

This reluctance is buttressed by the attitudes of industrial associations. Approval of any complementation scheme must be obtained through the *sindicatos* of the various industrial sectors. These syndicates are both local and national, with the local ones being much more powerful in most Latin American countries. Conflicts arise from different local viewpoints within an industrial syndicate; a continuing one arises in Brazil between Rio de Janeiro and São Paulo members. The composition of these organizations makes it difficult for agreement to be reached. Further, there is a built-in obstacle in that the officers of the industrial syndicates see a loss of their roles if regional integration occurs; in order to get members to support the organization, they need an issue, and opposition to regional integration provides just such an issue and protects the role of the executive officers. Some regional organizations, notably the Consejo Interamericano de Comercio y Produccion (CICYP), have strongly supported integration. But CICYP's membership cuts across industrial lines, and though it represents the larger, more aggressive, and progressive companies, its influence at the national level is not as strong as that of the industrial syndicates.

A pressure on private enterprises to support integration arises from their belief that it would limit the ability of governments to interfere in the process of economic growth. The necessity to agree regionally would divert energies and restrain the more revolutionary governments. But when it comes to active support as distinct from verbal encouragement, specific interests (and uncertainties) tend to overcome aggregate policy objectives of the industrial sector.

International Companies. Given the attitudes within Latin America, a few observers assert that an initiative for integration must be taken by the international companies. These companies are not in a position to push integration, however, nor are they always interested. They would like to have export markets, but these are sought *after* the domestic market is secured. And the existence of nationally oriented facilities becomes an obstacle (high cost of relocation or adjustment) to integration. In the main, investments by international companies in a particular market have not been made with regional integration in mind. Officials of nineteen international companies at a session during 1969 in Latin America could not think of an investment by any private company made to serve a regionally integrated market. At a meeting on a proposed complementation agreement, representatives of major international companies were instructed to keep imports out of the national market. Even among those companies that have similar or complementary plants in two or more countries in Latin America, the experience shows a variety of cost patterns, but none clear enough to tell the company how it would be best to integrate regionally if given the chance.

A major obstacle to the initiative being taken by any international company is the fact that all recognize that integration means rationalization of the industry to a smaller number of companies. None of them can afford to initiate the suggestion for fear of being asked to drop out. In Argentina, after a cut from twenty-two companies to nine, it has seemed clear that none of the remaining companies would offer to sell out. All want the opportunities provided by a continuing presence; if they drop out, their options for the future are narrowed significantly. If there are any to be dropped, the U.S. international companies are in the stronger positions to remain and take advantage of integration, for the European and Japanese companies are not in a sufficient number of countries to make integration of their affiliates as profitable.

Although the international companies would have welcomed a larger market in which to invest, their past investment would become less profitable if they had to face competition within their local market. Even so, most would be willing to make substantial adjustments in order to have the future larger market, but recent efforts on their part to encourage industrial integration have come to nought. Of course, there is some skepticism among government officials when a foreign-owned affiliate comes in with a company complementation agreement—who will benefit?

Some companies have spent substantial amounts of executive time, both in Latin America and in the United States, putting together appropriate integration schemes for the company or industry. But home offices are reluctant to commit more time until governments show greater receptivity; in reply, officials of the affiliates argue that the governments will be receptive only to a concrete proposal—not to a mere idea. Government officials have asserted that they will be willing to examine carefully any concrete proposal to determine the benefits and costs, and that they will make a judgment on the specific advantages. But they are not willing to declare themselves in favor of all complementation agreements or even to state the criteria of their decisions to approve; their response will depend on the individual case. This attitude makes it difficult for company officials to know how to decide among alternative routes to complementation, and much time is required in preparing, reviewing, negotiating, and renegotiating—not only with government officials but with others in the industry. (Also, as stated earlier, negotiation with other companies is virtually prohibited by U.S. antitrust laws.)

In the automobile industry, since the assemblers produce only engines (Argentina, Brazil, and Mexico) and some stampings (Brazil), complementation agreements would not be much to their advantage unless rationalization of models or components were possible. As a beginning, in order to get the principle established of intracompany specialization, they have pushed for bilateral arrangements, especially between Brazil and Argentina—to no avail. For example, the proposed specialization by model between these two countries has been hung up on the question of who will produce which models and the problem of administrative balancing of the trade volumes (prices and costs, volumes and value, use of export incentives, and profit allocations and remissions).

Some of the international companies would have some problems in adapting to integration. Models are sufficiently different to require investment changes, as in the case of the Argentine Ford and several of its accessories, which have been redesigned for use in that country. In addition, it is not easy to get affiliates to make the accommodations necessary to work with each other—they do not necessarily trust each other and do not want to be dependent on each other for components or models. The decision to integrate might have to be dictated from Detroit, after its own assessment of the advantages of dovetailing operations and difficulties of getting management cooperation plus the costs of investment.

However, some of the affiliates of the U.S. auto companies have tried, on their own, to establish bilateral exchanges among themselves in Latin America. Two problems besides governmental attitudes have made it difficult: one is the suitability of components from one affiliate in the models produced by the other and the other is the problem of pricing (for each considers the others' prices wholly fictitious). Some success has been achieved between affiliates of a U.S. company in Venezuela and Chile, with governmental approval. The same Venezuelan affiliate is moving to integrate with affiliates in Peru and Argentina, without benefit of tariff reductions but with governmental approvals, which seek to assure a bilateral balancing of trade volumes.

Although the international companies are permitting affiliates to work out bilateral exchanges on their own, if they can, they are not taking regional initiatives. If they did, it would probably be counterproductive, given the present attitudes in Latin America. In sum, the prospect for use of complementation agreements to integrate the key industries of petrochemicals and automobiles is bleak (and the same can probably be said of electronics). If positive moves can be made in other sectors, maybe their lead could be followed in the more critical industries, but that will take considerable time and the die may be cast too firmly by then.

If integration is to occur—as all say it should (or must)—new and different initiatives are necessary.

A Proposal

As previously observed, if there is to be effective integration, it will come about when it is apparent to all participants that the rewards are both high and certain. There are two ways of widening the opportunities to increase efficiency and still provide equity: (1) creating a mechanism for integrating several industrial sectors simultaneously within a region, or (2) increasing the number of participants to be included in the sectoral agreement so as to encompass the advanced countries. The first would lead to specialization by participants among industry sectors within the region; the second to international specialization within a sector. Neither would match exactly the specialization that would occur under free trade, but the perceived cost of specialization would be reduced for each country.

Multisector Agreements

In order to gain an internationally competitive level of efficiency, both the automobile and petrochemical sectors in Latin America perforce would have a skewed distribution of production facilities over the region. Some of the participants would gain more than others. The skewed pattern would result partly from (static) comparative advantages; i.e., from historical accident, highly nationalistic industrial policies of the past and the present distribution of market demand among the countries. Whether this distribution of facilities would be appropriate over a period of time, after changes in industrial structures and shifts in market growths, is a question bothering governments. Should they accept the present pattern rather than paying a present price to obtain future gains of a less-skewed distribution of facilities?

Important benefits from industrial integration—advanced technology, employment, and nil payments impacts—can be obtained without distributing various parts of all the major industries to each country. Advanced technology can be obtained in many different industries. And opportunities for scientists and high-level technicians can be provided without having all such industries in one country. But it is required that the country have some industry using advanced technology. Similarly, employment is not dependent on the existence of specific industries—but some industries. And payments balancing can be achieved without regard to which industries are being promoted.

The solution, therefore, lies in the allocation of facilities from among several industrial sectors simultaneously among the several participants.* If one country is squeezed out of participation in one section, it may be satisfied with a role in another. Since most industrial sectors lend themselves to some specialization among the production stages, the cost differentials in allocating such facilities are not likely to be large, given a sufficient number of sectors involved. A more efficient pattern of production and trade would result through allocation of several sectors simultaneously than from balancing benefits within a single sector, although resistance to shifts in particular facilities or lines of production would probably increase with the number of sectors involved.

An effort among the Central American Common Market countries to allocate entire industries to each participant, with each specializing entirely in a few sectors, has not proved singularly successful. Each has been unwilling to give up all participating key industries. They have also found it difficult to determine the balance of benefits where a tire industry, for example, is compared to a paper industry. It is unlikely in Latin America that entire industries would be permitted to be located in a given country, to the exclusion of others. But with the larger Latin American market this would not be necessary for efficiency. Scale of operations would permit specialization among operations and duplication of some facilities.

The proposal to form complementation agreements on several industrial sectors simultaneously runs into the objection that the balancing would be too complicated. One of the reasons that the national-and-common-list approach has failed is the complex problem of determining benefits across a wide range of duty reductions. Simultaneous negotiation of agreements in several industries might also produce highly complex trade-offs and balancing acts. But a major reason for failure of the duty-reduction approach was the uncertainty as to the distribution of potential benefits. Complementation agreements would produce greater certainty and could also reduce the complexity through a process of self-selection among the participants: those interested and seeing an advantage would stick with it while the others would opt out. The number of complex decisions thus would be reduced to a more manageable level. The larger the number of countries, the larger the number of industries that might have to be involved in order to achieve an acceptable balance, but the easier it would be to achieve efficiency with equity because of the size of the market and number of trade-offs available.

---

* This assumes, of course, that national-security considerations are not the final determinants of policy, which they may well be.

International Specialization

If the market served is to be expanded, and if some of the problems between advanced and developing countries are to be met, consideration
should be given to negotiation of complementation agreements between the
Latin American countries and the United States and Europe. Such an opportunity would make integration much more attractive to the larger Latin
American countries, which see LAFTA integration as a drain from their
economies to the poorer countries. The agreements would also provide for
access to the markets in the advanced countries in ways agreed upon by
the latter, reducing frictions and arguments over preference.

Support for such agreements would probably be found in several of the
advanced countries. Considerable discussion has been generated in Europe
over recommendations in the Colonna Plan for industrial integration of
the EC. That plan proposes an industrial policy directed initially to technologically advanced industries that could set the pattern for other industries as well. The Colonna approach would involve some allocation of
production but would rely more heavily, in the long run, on coordination
of procurement policies and on joint research-and-development programs.
The policies proposed in the Colonna Plan are seen partly as a means of
combatting the penetration of the U.S. multinational enterprises.

In order to eliminate the potential conflict among industrial groups and
over the role of multinational enterprises, the U.S. government should be
interested in the discussion of a means of ordering competition and guiding
integration of the major industries over the world. Even some of the traditional industries, such as textiles, would benefit from an agreement among
the major producing countries as to the future pattern of development,
since such agreement presumably would eliminate the continued threat of
a return to restrictive trade practices. If discriminatory techniques are to
be employed for reasons of national interests, they should be brought under intergovernmental cooperative arrangements and used for agreed-upon
purposes only. The United States seems to be moving in this direction on
an ad hoc basis in the agreement on textiles; there are similar proposals for
shoes and electronics.

The advanced countries have a large stake in the continued freeing of
trade, but it seems unlikely that free trade will even be approximated
through a frontal attack on tariff and nontariff barriers. Restrictions that
remain have been adopted for reasons other than import protection, though
they may have this effect. Nontariff barriers often are not so much protection against imports of specific products as they are promotion of local
industrial capabilities, e.g., governmental procurement practices. Before
they are likely to be removed, the problems giving rise to these tactics must

be met. The problems reflect the concerns mentioned earlier, such as autonomy, efficiency, growth, and equity; these same concerns arise over integration or penetration by the multinational enterprise. Such nontariff barriers are likely to be removed only by finding appropriate means of sharing the benefits of industrial advance.

The sharing of benefits is also at the base of the pressure for preferences from the advanced to the developing countries. Such preferences could be more readily negotiated and would be more effective under a sectional arrangement than under across-the-board reductions. The extension of preferences would then be melded into an agreement that guided industrial development so as to take advantage of the existence of the preferences. Without some such arrangement, the extension of preferences is likely to be useless and frustrating to the developing countries.

What is being proposed here is not an international cartel. It is more on the order of the European Coal and Steel Community or the United States-Canadian automobile agreement; both of these have taken into account the volume and location of production and the pattern of trade in an effort to move as close as possible to long-run comparative advantage while equitably distributing the benefits. A first basis for discussion should be a careful sector-by-sector analysis of the economics of location within each industrial sector so as to determine the costs of a more equitable distribution of facilities and the resulting benefits. Investigation is likely to show that many activities within industrial sectors could be spread around the world at not too great a cost penalty compared to least-cost operations.

So long as labor is the major immobile factor, people are likely to place their national interests over increases in international welfare. An approximation of international equity will have to be achieved through moving economic activities among nations rather than moving people. If such movements must be made in order to advance the developing countries toward a competitive place in the world economy, it would be better (i.e., least costly) to make some arrangements covering the international markets; others might be appropriately restricted to specific regions.

The existence of the international companies provides both a vehicle for implementing international agreements and a reason for negotiating them. Without their technology, capital, and managerial skills, high levels of efficiency and low costs cannot be achieved in the developing countries (save for a few countries or over a period of several decades); the gaps between the advanced and developing countries are likely to widen in several sectors. To put the power of the multinational enterprises under the control of intergovernmental arrangements sector by sector would permit maximum use of its capabilities while removing the fears and concerns of governments, for provision would be made *beforehand* for the equitable sharing of benefits. Intergovernmental agreements to guide the location of industrial activity and the utilization of the world's resources within selected sectors, so as to achieve both efficiency and equity, would avoid the twin pitfalls of a return to generalized trade barriers and of restrictions on the flow of capital and direct investment.

Appendix A
The Automotive Industry

Latin American governments have induced investment by foreign automotive companies without sufficiently limiting their number, ending up with more companies than the markets can sustain. Given the existing duplication of facilities, regional integration will require substantial shifts in the location of production. Trade-offs will have to be made relative to the structure of national production, ownership, import and export patterns, and prices.

Present Structures

Local investment in assembly of automobiles and trucks was induced by imposition of high tariffs on the finished vehicles. Duties as low as 10 per cent on the final vehicles provided 200 per cent protection on the final assembly process, since final assembly amounted to only about 5 per cent of the final cost. Completely knocked-down units could be imported without the duty and could be assembled virtually in garages, requiring little capital outlay. As a consequence, sometimes more than twenty makes of cars were assembled in one country.

Governments then added the requirement that locally made parts be used in assembly. Local operations became more complex and required extensive capital investment; the engine block, for example, would be imported, but the pistons, fans, and so forth would be locally produced. The capital commitment of the countries increased and the industry structure became more rigid, though the number of makes of automobiles may have dropped from twenty to ten. The major companies are mostly assemblers, sometimes called "terminal companies," buying a large percent of the final value of the vehicles either from local suppliers or the foreign parent companies.

National Production

The structure of the industry within each country has been largely determined by the nature of the content requirements, the response of the international auto and truck producers to these requirements, and the size of the national market. In 1956 *Brazil* became the first to try to create a nationally integrated automotive industry; it set local-content requirements at 35 per cent for heavy trucks and buses and 50 per cent for passenger cars, with

---

The author of this paper, James Fox, was a Peace Corps volunteer and is now with the Agency for International Development in Costa Rica.

light trucks in between. The requirements were raised to 90 and 95 per cent in 1960 and to 98 and 100 per cent in 1961; preferential duties were provided for the remaining imported parts. Seventeen companies responded and eleven were permitted to enter; the government cut their production plans from 530,000 for the four-year period 1957–1960 to only 345,000 vehicles. This target was met: annual production in 1960 was 133,000 units rising to 350,000 in 1969. Roughly two-thirds were passenger cars with the rest commercial vehicles.

The Brazilian automotive industry accounts for some 7 per cent of gross national product, with sales of over $1 billion (equivalent) in 1968 and employs over 200,000 workers, 30 per cent in assembly companies and the rest in suppliers. The industry is concentrated largely in the São Paulo area. Practically every category of vehicle is produced, except tractors for tractor-trailers, but the range of choice is limited by the fact that there are only a few companies producing in each category. VW has a corner on the small-car market; Ford and Chrysler produce the luxury cars. Styling is similar to the foreign models (save for a Willys model designed for Brazil), but the models are not changed yearly.

As to share of the market, VW held 50 per cent in 1969, with the merged Ford Willys accounting for 25 per cent, GM for 15 per cent, Mercedes-Benz for 6 per cent, and Chrysler for 3 per cent. Toyota and Scania-Vabis produced some specialized vehicles, accounting for about 1 per cent, and Alfa Romeo entered in 1968, buying a small state-owned company.

Argentina imposed a similar local-content requirement in 1958; each company entering had to submit a plan to achieve 90 per cent local production by 1963; import licenses were required for components not produced in Argentina. Twenty-two companies had begun assembly by 1960, but many of them did not comply with the regulations; in 1962, five were refused permission to continue and eight others were placed under special inspection procedures. By 1966, only ten companies remained.

Production rose rapidly to a level of 166,000 vehicles in 1964, and more gradually to 218,000 in 1969, of which 60 per cent were passenger cars. As with Brazil, the faster growth has been in autos rather than trucks, which initially were several times that of auto production. FIAT has captured the small-car market in Argentina producing 25 per cent of all vehicles; VW has not entered. Ford accounted for nearly 25 per cent, GM about 14 per cent, IKA-Renault 14, Chrysler and Citroën about 8 per cent each, Peugeot S.A. 10 per cent, with Mercedes-Benz and IME (the government-owned truck producer) accounting for about 6 per cent together. The seven large companies produced between 15,000 and 50,000 vehicles each in 1969. Capacity utilization averaged 62 per cent in 1968.

This excess capacity is accentuated by the fact that a complete range of cars and commercial vehicles is produced in Argentina, and the number of models in each category is large. Only the largest-selling models will reach production levels of 20,000 per year. To reduce costs, model changes are made only after several years; for example, the 1969 models were the first change in most autos since 1964.

Mexico followed suit in 1962 with a program permitting import licenses for components only to firms achieving 60 per cent local content by September 1964, with production of the engine being a compulsory part of local content. Several companies having assembly facilities withdrew; eight made the investment to produce engines. By 1967, some $1 billion was invested by assemblers and supplier companies, providing employment for some 75,000 persons, two-thirds of whom are in the supplier firms. Seven of the eight assemblers are located in the Mexico City area, as are most of the suppliers.

The Mexican government controls the level of production by establishing quotas for each company, but the quotas can be exceeded if the assembler will export (his own or a supplier's production), or increase local content, or cut prices of final products. Each method provides for a different increase in the output permitted. The basic quotas established in 1965 have remained largely unchanged, but the smaller companies have received some increases. Despite efforts to utilize their quota limits more efficiently and to employ the incentive schemes, the percentage of the market held by the larger companies has dropped from seventy in 1964 to sixty in 1968.

Total production of vehicles had risen by 1969 to 160,000, of which over two-thirds were passenger cars. Chrysler and Ford held 45 per cent of the market, producing 35,000 units each; GM was close behind at 17 per cent; VW was still further back, holding 15 per cent; Renault held less than 10 per cent, and Rambler-Willys some 7 per cent, with Nissan about 5 per cent. As in Argentina, the annual production by the major companies ranged from 10,000 to 35,000 units, composed of a wide variety of models. Mexican model changes are made annually as in the United States, since the large import-content permits the redesigned components to be imported and the unchanged ones to be supplied in Mexico. The largest item of imports is body stampings, equal to about 20 per cent of the value of components. In 1965, the government tried to get stampings done locally, and would have cut the potential 60 per cent cost penalty by forcing one body style on all producers; however, all companies protested vigorously and successfully.

Chile began a local-content program in 1962, prohibiting imports of new cars. At that time there were twenty firms assembling vehicles in the country. Four more entered upon publication of the decree shutting off imports progressively. The decree also required that assembly be undertaken at Arica, just south of the Peruvian border and 1,000 miles north of the major market for vehicles. By 1967, the uneconomic impacts of this location became manifest to the government and it permitted removal to other locations; by 1969, it was actively encouraging relocation to the central zone of the country.

By 1966, the requirement to shift to Chilean (or Argentinian) parts in 45 per cent of the vehicle caused some companies to drop out; local suppliers of components simply could not meet production schedules. The requirement was increased to 53 per cent by 1968, 58 per cent in 1969, and 70 per cent in 1970. Only nine automobile companies remained. Production in 1968 had reached a total of only 18,000 vehicles with FIAT the largest at only 4,000, followed by Citroën, GM, and Peugeot with 2,000 to 3,000 each. Ford was in two companies, producing a combined output of 3,000 units. Eleven different models were produced in 1969 with production at 22,000 units.

In late 1970, President Allende announced the intention to reduce the number of companies to three and to eject most of the big foreign producers. He had been in conversation with Nissan—not presently in Chile —over the possibility of producing cheap cars and utility vehicles rather than the larger models presently produced. The government intends, apparently, to open contracts for auto assembly to international bidding, which would require the winners to set up joint ventures with the Chilean government holding a majority.

Venezuela started its local-content program in 1963 and required some specific items to be included. Sixteen companies responded, eight producing cars and eight assembling trucks. New companies were precluded from entering by a 1965 decree, and a programing group was established for the industry with the power to set local-content percentages and require specific items to be included as soon as their quantity and quality became adequate (regardless of price). In 1968, the local content had to be 34.5 per cent by weight, was raised to 38.5 per cent in 1969, and was expected to go to 41 per cent. The "by weight" requirement means that the lower-valued, heavier components are produced in Venezuela and the lighter, more complex, and valuable parts are imported. This division of production is reinforced by the decree prohibiting assemblers from producing components; they are normally more interested in the higher-valued components themselves. The components produced locally include generators, tires, glass, radiators, paints, upholstery, wheels, frames, and some springs. Engines, transmissions, axles, and body parts are wholly imported.

Production in 1968 had risen to 63,000 vehicles, with 7,000 persons employed, largely in the central zone of Caracas and Valencia. Ford and GM are the largest producers, with each selling 15,000 units; Chrysler sold 10,000, VW 6,000, FIAT 3,500, and Chrysler 3,000. Eight others aggregated only 7,500.

Peru inaugurated its program only in 1965, though its decree covering local content was issued in 1963. The decree provided preferential duties on imports of components and knocked-down units if assemblers reached a 30 per cent local content in five years. Fully assembled cars could be imported but only at the higher rate; however, in 1967, imports of luxury (over $3,000) passenger cars were prohibited. By 1969, thirteen companies had established assembly plants, raising output from 3,000 units in 1965 to 17,000 in 1968 but only 10,000 in 1969. Ford was the largest in 1968 at 7,500, with GM at 4,000, Chrysler at 2,000, and VW at 1,000. In 1970, the new government announced intentions to reduce the number of companies to four, either through merger or a move such as that taken by Chile; conversations were begun with Japanese interests, including automobile companies.

Colombia still imports most of its vehicle requirements, but out of some 20,000 units sold, a total of 3,000 were assembled by an affiliate of Chrysler and one held by Willys and Peugeot jointly with local interests. The assemblers do get some duty reduction, but they must use locally-manufactured parts "to the extent that they can be obtained" in adequate quantity and quality—meaning between 35 and 40 per cent. The price of such items may be as much as ten times the price of imports from the U.S. Local content is supposed to rise to 70 per cent within eight to ten years.

In an effort to restructure the Colombian industry, the government opened a competition for the right to establish new car and truck assembly plants, with the stipulation that the proposal provide for 50 per cent ownership by the Instituto Fomento Industrial, the government-sponsored industrial development agency. Twenty companies submitted bids, but some rejected the ownership requirement. Renault (owned by the French government) was selected for the passenger car plant; the Mexican government company, Diesel Nacional, (known as DINA), was selected for truck assembly, but it finally dropped out, for reasons not made public. The capacity of the Renault plant is set at 15,000 vehicles, beginning production in 1971, and using 25 per cent local content.

Uruguay has also attempted to promote local assembly by similar requirements, prohibiting imports of fully assembled vehicles and of certain components that are available locally. Although some twenty companies have operations in the country, total sales have averaged only about 3,000 units during the late 1960's and dropped under 1,000 in 1968.

None of the remaining countries—Paraguay, Ecuador, and Bolivia— has any domestic automotive industry; total demand during 1968 amounted to under 15,000 units.

In sum, total Latin American output of vehicles in 1969 amounted to about 770,000 units, roughly two-thirds of which were passenger cars. Brazil produced nearly half the total, with over one-fourth in Argentina, and over 20 per cent in Mexico, leaving less than 10 per cent for the others.

Even within the major producers, sales were widely distributed among competing companies and even more numerous models, so that production runs were under 50,000 units annually, save for the VW in Brazil, which topped 150,000. The impact of local-content requirement was heaviest in the largest country, Brazil, which could sustain larger production runs among suppliers.

Ownership

If a restructuring of the industry is to occur across Latin America, the existing pattern of ownership will be of singular importance. If the companies assembling and producing components were all owned by nationals or by Latin Americans, mergers and concentrations would be easier. But the assemblers are owned by international companies and the suppliers largely by locals or other foreign companies. Nor is the pattern of international ownership repeated from one country to another. The three major U.S. companies are in all Latin American countries which produce automobiles; they generally own 100 per cent of the affiliates, as do all other companies. (See Table A-1). Only Chrysler had two minority holdings. No other major international company is in all producing countries.

Market penetration by each company varies widely. VW, for example, is the dominant company in Brazil, but does not exist in Argentina, ranks a close fourth in Mexico, a distant fourth in Venezuela, and has no operations in Peru. Ford bought out Willys in Brazil to rank second in that country; it ranks second to FIAT in Argentina, alternates with Chrysler for first place in Mexico, ranks fifth in Chile, but was not expected to survive the cut in Peru. FIAT is the largest company in Argentina, but has operations nowhere else except Chile, where it is also the largest.

GM ranks third in Brazil, fourth in Argentina, third in Mexico and Chile, and probably first in Peru. Chrysler is fourth in Brazil, sixth in Argentina, tied for first in Mexico, has a minority interest in a small Chilean company, and ranks third in Peru. Renault is not represented in Brazil but is third in Argentina, fifth in Mexico, and hardly shows in Chile or Peru. The other companies—European and Japanese plus American Motors—are in only one or two countries. Thus, only the three major American companies are in all the major countries and with sufficient existing capacity to make integration appealing to them if barriers were removed. However, both Chile and Peru seem to be leaning heavily toward Japanese companies for their plans to restructure the industry.

Few of these major assemblers share ownership of the enterprise with local interests. However, Ford owns only 51 per cent of Ford Willys in Brazil, as a result of having bought the Willys operation that had issued

Table A-1
Location and Ownership of Latin American Automobile Companies, 1969

Owning Company	Percentage Equity in Affiliate					
	Brazil	Argentina	Mexico	Chile[a]	Peru[b]	Venezuela[c]
Ford	51	100	100	100	100	100
General Motors	100	100	100	100	100	100
Chrysler	92	99	33	33	100	100
American Motors	—	10	10	—	—	—
Volkswagen	80	—	100	—	100	100
Fiat	—	100	—	100	100	100
Renault	—	—	(lic.)	100	—	(lic.)
Peugeot	—	100	—	(lic.)	—	—
Daimler-Benz	51	91	—	—	—	?
Toyota	100	—	—	—	100	100
Nissan	—	—	100	—	75	100
Volvo (trucks)	100	—	—	—	100	—
Alfa Romeo	100	—	—	—	—	—
Citroën	—	100	—	100	—	—
Local Company:						
IME	—	(govt.)	—	—	—	—
DINA	—	—	(govt., Renault licensee)	—	—	—
FANASA	—	—	(govt.)	—	—	—

[a] Also, Ford holds 50 per cent of Chilemotores; GM has a licensee; Skoda owns 100 per cent of a company; Leyland owns a majority of a company; and NSU has a licensee.

[b] Plus licensee of Mack and Rover; a minority interest is held by International Harvester; a company owned 100 per cent by Rootes was taken over by Chrysler when it bought Rootes in the United Kingdom.

[c] Plus 100 per cent-owned companies of International Harvester, Isuzu, Leyland, and licensee of Renault and Rambler.

Sources: Various government and international publications, plus interviews.

shares locally. VW has a 20 per cent local interest in Brazil, and Chrysler an 8 per cent local shareholding, but GM's affiliate is wholly owned. In Argentina, all are wholly owned save the IKA-Renault company, which is majority held locally and shared otherwise between American Motors and Renault. In Mexico, all major companies are wholly owned except for Chrysler's minority position in Automex. Chrysler again is the only significant exception to 100 per cent ownership in Chile. And in Peru all were 100 per cent owned.

Even among the supplier companies, 100 per cent ownership by foreign companies is widespread. In *Brazil,* Clark Equipment Company, Timken

Company, Borg-Warner Corporation, Champion, and Eaton Yale & Towne Inc. hold wholly owned affiliates, with Eaton Yale & Towne holding 90 per cent of a second affiliate. TRW Incorporated holds three separate affiliates by 70 per cent, 60 per cent, and a minority share; North American Rockwell holds three minority affiliates; and Dana Corporation holds one at 75 per cent. In *Argentina,* Ford, Eaton Yale & Towne, TRW, Budd Company, and Bendix Corporation hold wholly owned suppliers, with Dana Corporation and Associated Spring Corporation holding minority affiliates. There is no rule against assembly companies owning suppliers, so FIAT is nearly 100 per cent integrated; other assemblers also own some suppliers, but the government was beginning in 1969 to discourage further vertical integration. In *Mexico,* only Bendix has a 100 per cent-owned affiliate: Borg-Warner holds one 55 per cent, but North American Rockwell, Budd, Eaton Yale & Towne, Clark, and TRW hold only minority shares. In *Chile,* North American Rockwell is the only visible U.S. company, but two Argentine companies have established affiliates. In Venezuela, ownership of suppliers by assembly companies is prohibited by law. Little is known of the European companies that may own affiliates in the various automobile sectors. But it is apparent that by far the largest number of local suppliers are locally owned. The only evidence of the distribution of ownership among Brazilian supplier companies comes from a report that investment by 1963 amounted to $800 million, of which 37 per cent was Brazilian, 33 per cent American, and 18 per cent German; the heavy investment by the foreigners was in assembly plants.

The American suppliers, though not numerous, held key positions in the industry; in many instances they were the sole suppliers of the components they produced: Bendix for brakes, TRW for steering gears, and Eaton Yale & Towne for valves in Brazil, for example. In Argentina, American companies bought out existing suppliers, establishing themselves as sole suppliers in some instances. But in Mexico, the government has insisted that U.S. companies sell up to 60 per cent of the shares of affiliates to Mexicans; even where there are sole suppliers, they are more Mexican than foreign owned.

Mergers have already occurred among the assembly companies, with the result that the foreign-owned companies have become larger. In *Brazil,* Vemag was taken over by VW; and Willys (already foreign owned with local participation) by Ford. International Harvester Company also sold its truck plant to Chrysler after substantial losses. In Argentina, all locally owned companies were gradually eliminated, save for IKA-Renault, which has substantial public shares outstanding, and the government-owned IME. But IME stopped auto production in 1964 and specialized in a single model of diesel-powered, pickup truck. IKA was formed by Kaiser and continued after the parent company folded in the U.S.; without parental assistance in design and technology, the company began to fail; Renault acquired a large shareholding in 1967. No mergers have occurred in *Mexico,* but FANASA, owned by Borgward of Germany, defaulted on loans in 1969 and was taken over by the government. It was to be merged with DINA and Vamsa into Automex (Chrysler), the last two having only minority foreign shareholdings; but negotiations failed, and it appeared that DINA would take over the FANASA facilities. Though no mergers have occurred in *Chile,* the foreign enterprises have gradually increased their penetration by buying into their licensees; GM, Ford, Chrysler, and Leyland took this route.

In sum, the ownership of the Latin American auto industry is largely in the hands of foreign companies even if both assemblers and suppliers are counted, with the assemblers all but completely foreign owned. Further efforts to rationalize the industry or to integrate it regionally will likely mean an increasing concentration of output in a few companies. And rationalization of the supplier companies would give a greater role to those foreign-owned suppliers located in several countries.

Vertical Integration

The size of the market in each Latin American country is so small that suppliers tend to serve the entire set of assembly companies; they are either sole suppliers or there are a few competitors serving all customers. This fact has tended to reduce vertical integration within the industry. In Brazil, the largest market by far, purchases from independent suppliers amounted to 63 per cent of sales by assemblers in 1967. VW bought 54 per cent but Toyota 80 per cent from outside companies. The items normally produced by the assembly (or terminal) companies include the more intricately machined items—engine blocks and cylinder heads—and those requiring greater capital outlays, such as body stampings; next in line would be transmissions and axles, then generators.

In neither Brazil nor Argentina are there any limitations on backward integration by the terminal companies. But Argentina follows the Brazilian model, with a few suppliers selling to all companies, achieving economies of scale in this fashion (e.g., Transax produces transmissions and axles). All assemblers bought parts and materials equal to 58 per cent of sales in 1967. The items produced by the terminal companies were mainly engines and body stampings.

Mexico has enforced a separation of suppliers from the terminal companies by regulations prohibiting the latter from producing anything other than engines. All parts and other components must be obtained from independent supplier firms, or imported. This policy, coupled with that requiring 60 per cent Mexican enterprise, reserves a substantial portion of the industry for Mexican enterprise. However, the assemblers can still import 40 per cent of the final value of the vehicles.

Chile follows Brazil and Argentina in not limiting vertical integration; indeed, in order to raise local content it had to persuade the terminal companies to establish parts facilities. The tactic was to permit these companies to move from Arica to central Chile if they would invest in the supplier operations; the cost to the assemblers was between $3 million and $4 million. Here also, vertical integration is limited by the import of all parts permitted.

In sum, in order to spread the secondary and tertiary benefits of the automobile industry, a number of supplier companies have been created in each country. Their ties to the terminal companies have been loosened by the necessity to sell to all assemblers in order to achieve economies of scale.

Imports and Exports

Policies to create local industry have included strict limitations on imports of vehicles and parts; these limitations are usually by value, though some are by weight. The consequent desirability of obtaining imports from the least-cost source has meant that little intraregional trade has been stimulated within Latin America. Shutting out American and European imports also shuts out imports from other Latin American countries. Nor have these parts been the subject of effective concessions under the national lists. Exports of vehicles parts to countries outside of Latin America have been virtually precluded by the resulting high costs and prices of production in Latin America.

Imports into *Brazil* are limited to a "technically necessary" 1 per cent of the value of automobile production and 2 per cent of truck output. Items such as bearings fit this category and have amounted to approximately the legal limit. Import costs of auto production are much larger if materials going into supplier production are counted; for example, all steel used in body stampings is imported, as are aluminum and magnesium ingots and alloy steels. Further, the 100 per cent duty on fully assembled vehicles has not prevented all such imports; some $11 million was spent on importing 1,000 passenger cars in 1967. In addition, $45 million was spent on tractors for tractor-trailers (not produced in Brazil) and other special vehicles.

Exports from Brazil are of an even smaller volume, amounting to $3.2 million in 1965, and the government has taken little interest in trying to stimulate them. GM has exported some engine blocks to its South African affiliate, and Ford ships crankshafts and camshafts to Argentina. Other than a few hundred trucks and buses, mostly to Paraguay, these parts sales constitute the entire exports from Brazil.

Argentine imports of parts and components are limited to 5 per cent for passenger cars and 13 per cent for trucks; they amounted to $45 million in 1967, for an average percentage of 7 per cent of the value of total output. The value of the direct imports of parts is somewhat artificial, however, for a system of valuation based on weight of different categories of goods is used in order to prevent assemblers from underinvoicing imports. If imports going into locally produced components are counted, the proportion of imports in the final vehicle would probably rise to closer to 20 per cent on the average.

To increase the potentials for LAFTA trade in automobile parts, the Argentine government announced in 1965 its willingness to allow imports from any other country in Latin America to be counted as local content, provided that the other country bought a similar amount of automobile parts from Argentina. More recent decrees have limited the arrangement to 6 per cent of the total value of local content and to 10 per cent of the local content of any one model produced, whichever is smaller. Further, such imports cannot account for more than 30 per cent of the firm's purchases of the part or component if it is also made by other Argentine suppliers. A nominal duty of 5 per cent is charged on such imports.

The amount of trade with Chile under this arrangement rose to $7.3 million in 1968. Argentine exports were largely engines for Chilean affiliates of companies also in Argentina; those from Chile cover the entire range of its parts production. To facilitate such exchanges, Ford obtained an agreement from Argentine and Chilean railroads to allow direct shipments of Ford products between Buenos Aires and Santiago without transfer at the border. The arrangement cut transport costs by 40 per cent and time from several weeks by ship to one week by rail.

Mexican imports of parts and components amounted to some $200 million in 1968, mostly body stampings; but steering gears, wheels, axles, and transmission parts were also included, depending on local availability. The volume of imports amounted to about 40 per cent value of total output— the maximum allowed under governmental regulations. These items came almost wholly from the U.S., German, and other parent companies (or their suppliers). It would not pay the assemblers to import from other Latin American countries, substituting for local-content items, since Mexican costs are generally lower.

To stimulate exports, Mexico altered its regulations on production quotas to require that, beginning in 1970, firms would be required to export an amount equal to 5 per cent of its imports or have its quota cut by 5 per cent. For 1971, the export requirement is to be increased to 15 per cent of imports, and exports are supposed to expand in subsequent years until they match imports. The foreign-owned companies found little difficulty in meeting the 1970 requirement, since they were already exporting 5 per cent of imports, but the government-owned companies do not have foreign ties to help obtain export markets; they will find it difficult to comply.

Foreign-owned companies increased exports because they could thereby raise their production quotas. The companies responding were Ford, GM, Nissan, Chrysler, and VW—each having foreign parents who were able to open up exports to sister affiliates in third countries. The government has urged VW to export components to the United States (where it does not have an assembly or production line), but Mexican costs made the effort uneconomic. VW tried to comply but settled for exporting Mexican parts for inventories in the United States. Chrysler and Nissan have exported engines to Chilean affiliates and sourced some smaller parts, such as radios and radiators, in Chile in order to take advantage of Chilean government regulations. The major part of these exports, seem to be unrelated to incentives in other countries, responding only to Mexican incentives. Ford, GM, and Chrysler have all begun sourcing some engine requirements for Venezuelan affiliates from their Mexican production. Ford has shipped tooling and engine parts to the United States and to European affiliates, and under an arrangement between the governments, Ford engines will be exported to Chile in exchange for coil springs (limited to 20 per cent of Mexican productions).

The total amount of trade in all automobile parts was $13 million in 1968, but the 1969 exports were nearly double that level. In the first major export sale involving a Mexican-owned supplier firm, Ford (United States) has contracted to buy 400,000 transmissions, valued at $39 million, over four years from Ford (Mexico), which gets the credit for the export and an expansion in its production quota. Exports by Mexican suppliers are never made directly but only through an assembly company, which in turn sells to its affiliates (usually the parent).

Mexican government officials and more than a hundred industry representatives went to Brazil to work out a reciprocal trade package; over fifty deals were agreed upon, but no real integration resulted. Each agreed to reduce duties on items exchanged to 5 per cent (Brazil from as high as 65 per cent and Mexico from 105 per cent). Such exchanges are made difficult by a Mexican policy of prohibiting imports if local production of any *new* item can be achieved at costs no greater than 25 per cent above imports.

Chile's regulations permit imports to be counted as local content if a similar amount of Chilean parts are exported; up to 20 per cent of the 58 per cent domestic content can be so imported. Roughly 42 per cent comes from the United States for an American-style auto, 40 per cent from Chilean sources, and 18 per cent from Argentina. Total imports amounted to nearly $8.5 million in 1968, with exports just over $5 million; the exports went solely to Argentina, but some imports came also from Mexico. Ford and Peugeot failed to equalize exports with imports in that year and were given twelve months to balance them or cease operations. Nissan was forced to suspend operations in 1967 for failure to balance its LAFTA exports and imports. Although Chile produces radiators, steering wheels, springs, heaters, jacks, filters, and nuts and bolts, it has not found a ready market in other countries since its prices are high—twice those in Argentina, for example.

Despite this cost penalty, as mentioned earlier, a U.S. company in Argentina moved its entire spring production to Chile to be able to export its Argentine engines to that country to gain the benefits of some specialization, even though the Chilean cost was 1.9 times that in Argentina. (Prior to the agreement, the Chilean differential was 6 to 7 times the Argentine price.) The exchange is at Argentine prices, with Chilean affiliates absorbing the cost differential. Another company has been induced to add truck production, which will use engines from Mexico in exchange for steering gear boxes from Chile. The Chilean cost is 3 times that in the United States, and the Mexican affiliate does not like this added cost; bilateral discussions between the affiliates were conducted at artificial (or "bargaining") prices, but the scale of operations in Mexico would permit that affiliate to buy ten $50 steering boxes for every $500–600 engine needed in Chile, making for a balanced exchange. A major problem in any such extensive exchange is the inability of either partner to be able to rely on the other continuing the arrangement. Even so, Chile has urged extensive compensation agreements with *all* other Latin American countries.

Venezuela is gradually opening up its export opportunities by substituting imports from other Latin American countries for items formerly imported from the United States or produced locally. For example, one U.S. assembler obtained agreement from the Chilean government to permit it to sell frames to Chile in exchange for coil springs, wheels, and bumpers. In order not to cut local production of the imported items, the assembler exports Venezuelan production of these same items to its Mexican affiliate; Mexico has had little production of coil springs. The inability of the affiliate to produce the parent company's own line of accessories and components prevents marketing such items through the worldwide network of affiliates. The same company hopes to make similar bilateral arrangements with affiliates in Argentina and Peru. None of these arrangements have had the benefit of tariff reductions. The governments have agreed, in order to obtain the benefits of balanced trade; tariff reductions would not alter the level of the exchange but might alter the relative prices and therefore change the precise items traded. Under existing price patterns, Venezuela could export frames for passenger cars to Mexico (which does not make them) in exchange for frames for commercial vehicles (not produced in Venezuela).

Some interesting trade shifts are observable under the complex government regulations as to trade and local content. For example, one assembler

having affiliates in Venezuela, Argentina, and Chile has worked out a tri-
lateral deal to maximize its benefits from the regulations. It exports Argentine engines to Chile, obtaining the local-content credit under the bilateral
agreement; reexports the engines to Venezuela, gaining the credit under
the trade arrangement between Chile and Venezuela and permitting the
Venezuela affiliate to export to Chile; Chile closes the circuit by parts exports to Argentina. Under such arrangements, it sometimes becomes profitable to *give* parts from an affiliate to another in order to gain the benefits
of the various incentives and economies of scale from larger production
volume.

Colombia accepts Chilean auto parts as local content, and Chile agreed
to import Renault engines from Colombia in exchange for Renault transmissions produced in Chile; the exchange is to be at international prices
for each—roughly two motors for one gearbox—with the cost penalties
from inefficiency being absorbed by each producer. The existing companies
are permitted to buy from Chile as local content but can import only items
not produced in Colombia and must, of course, export an equal value;
formerly, the export was not restricted to auto parts, but such a restriction
now exists.

Prices

A major problem in generating greater intra-LAFTA trade is the high prices
for parts and components produced locally. These high prices are reflected
in the substantially higher prices for completed vehicles in each country.

In *Brazil,* a VW sedan retails for $2,650 (equivalent) compared to
$1,295 in Germany. The compact automobiles produced by Ford and GM,
while not exactly comparable to cars they produce elsewhere, retail around
$4,000, which is considerably higher than automobiles of similar size in
other countries. This higher price is attributable largely to higher production costs but also to two taxes—a value-added tax of 17 per cent and an
excise tax ranging from 18 per cent to 24 per cent on passenger cars.

Argentine prices range from 2 to 2½ times world levels for the same
types of automobiles. A study by the Argentine automobile manufacturers
association found that the average excess of vehicle prices over international levels was 122 per cent in 1968. Of this excess, roughly 56 per cent
was due to the small-scale operation of the industry; 20 per cent to high
costs of changing dies, even if this is only done every three to four years;
the remaining 45 per cent to raw materials prices, high costs of imports,
taxation, and financing costs. A carburetor costs 3.5 times the U.S. price,
some castings and forgings up to 5 times (because of high cost of imported

pig iron), and axles up to 2 times. The elevation of prices has been held down somewhat by introduction of competition among suppliers; now only about half of the locally produced items are made by monopoly companies. The lack of price control frequently puts price determination in a bilateral-monopoly setting.

To push down prices in the industry, the government made an arrangement with FIAT in 1969 permitting free importation of capital equipment to go with a $100 million investment program if FIAT would cut prices by 30 per cent (or to a level 1.5 times Italian prices at that date) during the next five years. A second move was aimed at cutting supplier prices; it permits imports to be counted as local content if the cost (including duty and sales tax) is lower than prices of domestically produced parts. In addition, in mid-1970, a duty on fully assembled automobiles was set at 140 per cent in lieu of prohibition, and provision was made to cut it gradually to 70 per cent over a five-year period. Given the present prices of 100 to 150 per cent above international prices, there will be a downward pressure in the future, probably leading to further failures among the Argentine companies. In order to have some ability to prevent suppliers from raising prices, some of the assembling companies try to exceed local-content requirements by 1 or 2 per cent; they can then substitute imports (or threaten to do so) if a supplier raises prices. The significance of this threat is evidenced partly by the estimate that a reduction of local content by 5 percentage points would cut 30 percentage points off the average cost override of 122 per cent per vehicle; another 5 point drop would cut another 20 points off the override —at an output level of 12,500 units per company.

Mexican prices are controlled by the government at no higher than 60 per cent above world levels. Despite complaints from the industry that a profitable return on investment is not permitted by these levels, the government has refused to raise them. The government holds this position partly because it provides no higher protection for local suppliers than 60 per cent duties on existing producers and 25 per cent on newly established production. But these limits are effective for terminal companies *only* if they do not have to turn to higher-priced suppliers in order to meet the 60 per cent local-content requirement. In 1968, the cost penalties for Mexican-made parts over imported prices (ex duty) averaged 73.5 per cent (mostly engines plus axles, brakes, and transmissions that are required to be purchased locally). The cost penalties for some individual parts were several times international levels; the differentials range from 10 per cent to 1,500 per cent on some of the smaller items; but not all items can be compared for they are not the same as in the United States. The major reasons for the cost penalty are the high prices of raw materials to supplier companies, their small production volume, and labor inefficiency. In addition, most suppliers hold a monopoly in their product, which is supported by government policy and the high cost of machinery (e.g., for axles, brakes, and transmissions).

Vehicle prices in *Chile* are the highest in Latin America, averaging three to four times world levels; for example, a Ford Falcon (six cylinders) was $9,340 compared to the U.S. price of $2,333 in 1969; an F-100 truck was $8,030 compared to $2,354 in the United States. The Chilean price includes duties on imported parts and an 8 per cent turnover tax, but the major differentials are the high prices of components. Bumpers, for example, were 300 per cent of U.S. prices; wheels 178 per cent; radiators 207 per cent; tires 462 per cent; and seat assemblies 308 per cent in 1969. (Chilean regulations permit officials to set a supplier's price 5 times that of the U.S. price regardless of his costs.) To these price differentials must be added the cost to assemblers of not having supplies available at appropriate times. An example was noted earlier, that of the company which had to air-freight 600 bumpers from the United States because of the cost of waiting for local supplies. One company in 1969 had eleven U.S. experts trying to obtain local supplies. The problem was made more acute by suppliers' unwillingness to increase investment in the face of governmental uncertainties and political instabilities; suppliers would rather accept present profits than risk more funds seeking greater profit. Disproportionately large numbers of production engineers are needed to reschedule production according to shortages of components. There seems to be no great help from expanding production runs, for a fivefold increase in output by one supplier (partly owned by an American company) cut prices only 10 per cent. Prices seem to go *up* as production expands, since the government has adopted a policy of creating (locally owned) monopoly suppliers in key items. Another company official estimated that engine production in Chile could be efficient if one model were to be made and exported all over the world, but the investment would amount to $30 million—obviously too great a risk.

In *Venezuela* vehicle prices average only 30 to 50 per cent above world prices, largely due to the low level of local content. Even so, the government began to control prices in 1969 in an effort to keep them from rising further.

Such high levels prevent any but the most strictly balanced of trade among the LAFTA countries in parts and components; otherwise, the importing country adds substantially to its international payments. Even with balanced trade, the final cost of the vehicle is not substantially reduced unless economies of scale can be achieved in supplier production. But substantial levels of exports are prevented by the strict limits on bilateral exchanges. The automobile industries remain in a circular bind: high prices will continue to prevent substantial trade, and the absence of export markets will continue to prevent economies of scale, giving rise to high prices.

Benefits of Integration

The gains from integration for the member countries are increased to the extent that (a) trade creation exceeds trade diversion, (b) the members were competitive prior to integration and potentially complementary afterwards, (c) duties on former imports among the partners were at high levels, and (d) members trading after integration are the principal suppliers and principal markets for others within the region.

The Latin American automobile industry reflects each of these situations in a high degree. First, imports by the two largest auto-producing countries (Brazil and Argentina) are quite small; taking all LAFTA consumption, only 25 per cent is met through imports of vehicles, components, and parts. The extent of potential trade diversion is, therefore, limited in comparison to the potential trade creation. Second, the national industries are directly competitive in view of the fact that they have all been established so as to produce the same end products and most of the parts; equally, there is substantial potential for specialization through complementary production. Third, the restriction of imports has amounted to virtual prohibition in many countries, with strict quotas on permissible imports; creation of an integrated industry and reduction of these barriers would undoubtedly enhance welfare for the region. Fourth, since the objective of integration is to create a regionally self-sufficient automobile industry, all countries exporting would be principal suppliers and all importing would be principal markets; diversion of trade to third suppliers or markets would be small.

In addition to the shifts to restructure the present situation, some dynamic effects would occur to increase the benefits from integration. They include increased efficiency from economies of scale and greater competition, greater certainty as to the opportunities for investment, and acceleration of technological improvements. These gains are assessed for an integrated automobile industry and are then examined as to their distribution under different groupings of the Latin American countries.

Economies from Integration

Economies from integration arise from an expanded scale of operations within plants or firms and from factors inside and outside of the industry, unrelated to scale of operations.

Scale Factors. The automobile industry worldwide is characterized by exceedingly large production units, combined into companies of even larger scale, with usually only a few dominant companies in each national industry.

The United States has only three companies that are considered sufficiently large to be efficient. Europe can be expected, says the president of FIAT, to cut its number of companies to half a dozen over the next decade. And the Japanese industry, though the second largest national grouping, is not yet considered by Japanese government officials to be fully competitive.

Economies from large-size companies involve gains in managerial and financial operations. Economies from large-size plants involve increases in productivity of labor, which can be more specialized, and of machines, which may be larger or more durable. For example, size of machines (and output) or durability of dies may rise faster than the cost. Labor specialization is permitted by the switch from small-scale batch production to flow production. Both labor and machines may be specialized with flow production; with large-scale output, neither will be idle.

The size of plant that achieves the optimum economies of scale is a function of the technology used and the combination of labor and capital. Given the high cost of labor in the United States and Europe, compared to that in Latin America, it is probable that the most efficient-size plant will be more capital intensive in the advanced countries than in Latin America. But the most economic-size plant for components varies from item to item. There are some 3,000 separate components in an auto—over 20,000 if each nut and bolt is counted separately—and the processes of manufacture are quite different when sophisticated machinery and costly dies are needed (for example, the casting and machining of the engine block) compared to processes for standardized, general purpose items available off the shelf. The materials range from iron and steel (15 per cent and 70 per cent of a vehicle by weight, respectively) to plastics, cotton, copper, lead, zinc, glass, rubber, and various chemical-based products (all of which comprise the last 15 per cent of total weight).

There are three stages of production: manufacture of specific components; subassembly into engines, transmissions, rear axles, bodies, and so forth, and final assembly of the vehicle. Not all of these stages are comprised within an integrated automobile company in the United States; 20 to 25 per cent of the vehicle comes from industries outside of the automotive industry (steel and chemical); another 25 to 40 per cent comes from suppliers of parts and components; the rest of the value added comes from the auto company itself. Automobile manufacturers turn to outside suppliers when the latter can achieve economies of scale by serving several companies simultaneously with high-technology items or, at the other extreme, when suppliers can use general purpose machinery that can be shifted from production of one item to another at low cost. These supplier firms need not be large sized; of the 4,200 suppliers to GM's Chevrolet in 1952, 75 per cent were firms with less than a hundred workers and only 5 per cent had more than a thousand workers. The larger companies are those that have become almost single-source suppliers of specialized components to the entire industry: Bendix, TRW, Borg-Warner, Eaton Yale & Towne, North American Rockwell, and Dana (all have affiliates in Latin America).

The final assembly operation is more labor-intensive than the manufacturing of components. This last stage usually takes about 45 per cent of the total labor input but only about 18 per cent of the capital inputs, whereas manufacture of the motor, chassis, and suspension require only 33 per cent of the labor and nearly 50 per cent of the capital. The body manufacturing takes the remaining 23 per cent of labor and 33 per cent of capital. Given the fact that the assembly-line machines are more general purpose than specific, it is possible to handle different models on the same assembly line, reducing the scale economies of enlarged production once a sufficient production volme is reached. The economies of scale in *assembly* operations are reached at 60,000 units and probably exhausted at 100,000 units—not necessarily all of the same model. Given the lower labor costs in Brazil and Mexico, one U.S. affiliate has reported that, with an annual volume of only 20,000 vehicles, assembly costs there exceeded U.S. levels by only 6 per cent. This was accomplished by substituting labor and general purpose tools for some of the few specialized tools (such as welding jigs) used in the United States. Such substitutions reduce the significance of economies of scale in assembly.

Similarly, foundry operations are not highly automated in the United States. It is also possible to substitute labor for equipment in many foundry processes, making the cost effects of scale in that operation about equal to those in assembly. In Latin America, plants producing castings are used (even when owned by the automobile companies) to supply products to competing car makers and suppliers, thus expanding scale. Ford and GM have their own foundries in Mexico and Brazil, but in Argentina, they buy these components from local suppliers. At a volume of production of 20,-000 units per year, the costs of production vary widely in Latin America for castings—from 30 per cent above U.S. costs in Brazil for those made by the automobile manufacturer to 80 per cent above for those bought from suppliers, and from 120 per cent in Mexico for those made by the companies themselves to 300 per cent in Argentina on items from suppliers. Since scale of operations is similar, cost differences are attributable to other factors—usually differentials in cost of raw materials or (in Argentina) heavy overhead resulting from overcapacity.

Significant economies of scale are obtained, however, in machining of the engine block, which is required to achieve the proper tolerances. In the United States, the optimum scale on an automated line is around 500,000 per year for a single engine. Machining facilities in Latin America are much less automated, and many adjustments must be made to the low volume. IKA-Renault in Argentina cast an extension on its four-cylinder engine to give them the same length as the six-cylinder so as not to have to change the machining line. Despite various changes, one U.S. company official estimated a cost of 50 per cent above U.S. engine prices at an annual volume of 15,000–20,000 units. An official of an Argentine affiliate stated that it would require only small investment and a small change or two in the setup of the line to double or triple his engine output. Another calculated that an efficient engine plant could be set up in Latin America for an investment of some $30 million or more. But companies do not like to make significant financial commitments without assurance that the regional market would be open to them. Cost comparisons in the United States at different volumes of production of engines indicates the significance of scale economies: Using data for Australia, one researcher found that unit costs dropped from a level of 100 per cent on 25,000 engines per year to 90 per cent at 50,000, to 85 per cent at 100,000 and to 75 per cent at 200,000, while capital investment rose to 135 per cent, 205 per cent, and 400 per cent for the respective volumes.

Output	Initial investment in tools and equipment	Production cost of engine	Tooling cost of engine
40 per hour— (128,000/year)	$40 million	$225	$ 31
1 per hour— (3,200/year)	8 million	400	250

Body stampings produce the greatest economies of scale in automobile manufacturing and require a million units per year to reach optimum levels. At lower volumes, production is in batches, with periodic changing of dies, which raises costs because of downtime. Tooling and die costs are fixed— related to planned scale. More presses are needed to produce a larger output, but they can be used with several different dies and over several years or models. Dies can be used over several years if the models are not changed. Thus, the cost of both is not directly correlated with scale of operations—rather with capacity. These high fixed costs have persuaded Latin American producers not to change models so readily. At an output of 20,000 units per year, body tooling costs drop from $8 million to $2 million for each subsequent year with no model change. Economies of scale cause the totaling costs of 100,000 units in the first year to rise only to $10 million and to $3.5 million in subsequent years. Therefore, with a three-year model cycle, per car tooling costs decline from $205 at 20,000 units to $58 for 100,000 units.

Because of the high cost of body stampings, U.S. auto producers in Mexico import these items from the United States, making possible annual model changes; but in Argentina and Brazil, where stampings must be produced locally, model changes are infrequent. In Argentina, the first model change did not occur until 1969, six to seven years after the initiation of production.

Transmissions and rear axles are usually produced by supplier firms, which serve several automobile and truck manufacturers. Scale of production is nearly as significant in transmissions as in body stampings, and likewise axles are nearly as sensitive to scale as machining of engines. This is due to the importance of technology in transmissions, the high capital intensity of some of the processes for both transmissions and axles, and the rigid quality control necessary. At the present time, transmissions and rear axles are manufactured only in Argentina, Brazil, and Mexico, with one company producing transmissions for all users (save in Brazil, where Chrysler produces its own). Rear axles have been commonized by GM, Ford, and Chrysler in Brazil on truck models so that North American Rockwell could lengthen its production runs and cut tooling costs. The cost overruns on locally made transmissions and rear axles ranged from 100 per cent to 200 per cent above U.S. costs, though not all of this differential is attributable to scale.

In sum, scale factors alone would account for over 50 per cent of the differential between U.S. and Latin American costs (not prices, which also include substantial taxes).

Nonscale Cost Differentials. The rest of the differential arises from nonscale costs, such as those for raw materials, capital, labor, learning time, and the impacts of monopoly among suppliers or governmental price setting and local-content restrictions.

Raw materials represent about 20 to 25 per cent of the ex-factory price of a car and are made up mostly of iron and steel. If a weighted-average of these costs were raised 20 per cent, final cost would rise 4 to 5 per cent. The differential for hot-rolled sheet steel in Brazil is only about 4 per cent over the U.S. level, but around 50 per cent in Mexico, Chile, and Argentina. The differentials in various other steels, lead, tire rayon, and rubber in Argentina range from 75 to 250 per cent of U.S. levels. For all raw materials in Argentina, the wholesale price of the auto is raised 8 per cent through domestic supply and another 6 per cent from imported materials (including duties and surcharges); raw materials costs average 60 per cent above U.S. levels. Similar differentials exist in other countries, though not necessarily of the same magnitude in each.

Capital costs in Latin America vary from country to country but are substantially higher than in the United States. Differences depend on the extent to which the automobile companies tap the U.S. or European capital markets or rely on local sources. Substantial differences exist for both short-term and long-term needs. And other differences arise among the companies depending on their reputations. With effective interest rates ranging from twice to seven times those in the United States for different types and terms of capital among the different countries, it is difficult to calculate the impact of capital costs. A mitigating factor, however, is that the fixed investment per unit of output is relatively smaller in the Latin American setting, given the greater labor intensity of operations. This is offset by the greater working needs to finance somewhat larger inventories, as a result of uncertain deliveries, and to finance both suppliers and customers.

Labor costs per man-hour are considerably lower than in the United States. Average earnings by Latin American automobile workers per year are between one-fourth (Brazil) and one-half (Venezuela) of the U.S. level and approximate the range found in the European industries. Although these cost differentials should cut the final costs in Latin America significantly, differences in productivity and labor intensity vary considerably from the U.S. levels. Comparisons of this cost factor are thus made difficult.

Related to productivity is the problem of learning time, which was especially important during the early period of compliance with local-content requirements. Any increases in local content raise this cost factor. Long learning time results in poor quality, high rejects of components made locally, and erratic supply. These problems are reflected also in frequent breakdowns, poor scheduling, and inelastic total supply arising from inflexible production schedules. As a result of these problems, initial costs of locally made parts ranged from two to ten times U.S. levels. Continued technical assistance and financing from the major companies has shortened the learning curve and brought supplier prices down. Over the longer run, with regional integration, costs arising from the process of technical transfers should be down to international levels.

Monopoly elements also exist within the Latin American auto industry. Although only a few companies comprise the terminal stage of assembly and final sale, monopoly effects are weak in final sales because governments are vigilant and tend to keep lids on retail prices; also important is the pressure of oligopolistic competition and the elasticity of total demand for autos. But in the supplier sector, monopolistic elements are quite characteristic. Demand for components is relatively inelastic; governments do not hold supplier prices down; and frequently there is only one supplier, who must be used because of local-content requirements. Monopoly elements do not exist in the more standardized parts (nuts and bolts) but they do in glass, electrical parts, and others specific to vehicles.

It is difficult to make a precise assessment of the impact of these various cost differentials without going into detailed comparisons of specific parts and supply conditions within each country. Some of these nonscale factors can be affected by the scale of operations of the final assembler if a demand increase is sufficiently large to give the assembler a stronger negotiating position vis-à-vis a supplier; he can then force a price reduction or induce a shift in technology so as to cut costs.

Putting the various cost factors together, the study of George Maxcy and Aubrey Silberstein of the British automobile industry concluded that there were substantial economies of scale as production moved up from the lower ranges of output.* But study of the Argentine industry showed a slower drop in unit costs. The data from each were as follows (the cost base is not the same for each):

| | Index of Unit Costs | |
Annual Volume	Great Britain	Argentina
1,000	230	188
50,000	138	142
100,000	117	124
200,000	105	109
400,000	100	100

The Argentine progression will not hold in countries where imports of the more-costly items are possible. But it is probably applicable to Brazil, where the local content is quite high. In any import-substitution program, auto assemblers will tend to buy locally those components with the smallest excess cost; the added cost of local-content requirements will therefore rise slowly at first and more steeply at successive increases in percentage. The rate of cost increases begins to rise significantly at about 70 per cent local content.

Comparing the impacts of scale, nonscale factors, and local-content requirements among the major Latin American producers at a volume of output at 20,000 units per year, it was found that the index of production cost in Brazil rose from 120 at 50 per cent local content to 158 at 100 per cent, that in Mexico from 140 at 50 per cent to 158 at 60 per cent (the maximum presently required), and that in Argentina from 170 at 50 per cent to 250 at 90 per cent. The striking difference between Argentina and Brazil is explainable in part by the fact that nonintegrated suppliers are required in Argentina and the scale of production is smaller there. Costs of nonintegrated suppliers are 300 per cent above U.S. levels while those of components produced by U.S. affiliates in Argentina are only 90 per cent above U.S. costs. In Brazil, the differential between costs of components produced by U.S. assemblers and U.S. costs is only 10 to 30 per cent while those of nonintegrated suppliers are 80 to 170 per cent above—still substantially less than in Argentina.

---

* George Maxcy and Aubrey Silberstein, *The Motor Industry* (London: George Allen & Unwin, 1959).

In sum, as would be expected in any country with a fair industrial capability, substantial increases in scale and increased productivity through learning will permit increasing the local content to high percentages without sharp increases in unit costs. At the low volume of 5,000 units per year, local assembly from CKD (completely knocked down) vehicles can be undertaken with only moderate increases in cost. Still at this level, substitution of local components for about 20 per cent of value can be accomplished at little cost disadvantage, using standardized equipment—tires, batteries, radiators, nuts and bolts, etc. A local content above 20 per cent will begin to raise costs substantially at this level of output. But an increase in scale of operations to 10,000 or 20,000 will allow import substitution up to about 40 per cent without raising costs sharply, and a volume of 50,000 will permit local content of 60 per cent with less than a 20 per cent rise in unit costs. If volume can be raised to 200,000 per year in a single company, substitution of 90 to 95 per cent of the value of the vehicle becomes possible at a cost penalty of no more than 20 to 30 per cent.

The policy conclusions to be drawn from this assessment of economies of scale and other cost factors are that Latin America can reduce the costs of vehicles substantially and possibly become competitive on an international basis—at least in some major components—if it can reduce the number of companies through merger or elimination and can permit the growth of those remaining through regional integration to levels of production of at least 200,000 units each. If a level of 400,000 units per company could be reached, there would be no cost penalty against U.S. costs.

Costs Under Integration

Projected costs under integration will depend heavily on the scales of operation that can be reached, and these in turn will depend on the projected level of demand. Demand for motor vehicles is a function of price, income, and the age structure of existing vehicles. (That is to say, annual purchases are affected by purchases in previous years). Projections of demand for the next decade may be based on the assumption that pent-up demand has already been met in the major countries—Argentina, Brazil, and Mexico—but probably not in the smaller countries. The major countries have now been producing automobiles long enough to have absorbed demand that was not met when imports were prohibited. Of course, sales will rise as income rises, or if prices can be reduced.

The reaction of buyers to price reductions over time is not easy to assess, but such reductions may be interpreted as increases in income over time and adjustments made in the calculations on income. Impacts of price reductions within a shorter space of time can be estimated by assuming incomes are fixed and examining the reactions of individuals to price changes. Studies of Latin American demands indicate that the reactions to price changes on autos are less strong than those in the more advanced countries; reactions to increases in incomes are less strong also.*

In order to project incomes, the GNP growth rates for 1960–67 were used as rough indicators of future probabilities. Assuming that prices of automobiles in each country did not change and that incomes in Latin America increased about 4 per cent per year, demand for the region would rise to 1.4 million units in 1975 (including both passenger cars and trucks, of which 80 per cent will be cars) and to 2.2 million in 1980. Assuming that prices could be reduced to international levels by 1980 and assuming a continuation of the same growth in incomes, demand would rise in that year to 2.9 million vehicles. (See Table A-2)

Even with sales of 2.9 million vehicles, only four countries would have a total demand over 200,000 units, and only three over 400,000. Under such conditions, and assuming closed national markets, there would have to be only one company in Venezuela, one in Argentina, and no more than two each in Brazil and Mexico—if they were to produce at international cost levels. Alternatively, if free trade existed throughout the entire regional market, seven companies could have outputs above 400,000 units each; and they could then be competitive with world prices. However, Japanese officials do not feel fully able to face world competition even with an annual output of over 3.5 million units divided among a half-dozen companies.

If the pattern of the United States, the United Kingdom, and West Germany were duplicated the Latin American market would be divided 30 per cent each to the two largest firms, with the others taking proportionately smaller amounts. But what would remain to be shared (40 per cent of 2.9 million units, or 1.2 million units) would imply that only three other companies could be large enough to be competitive.

Local-content requirements could rise to over 90 per cent only if the region were integrated and no more than seven companies were formed; but "local content" then means "regional content." National content could rise to 90 per cent and still retain international cost levels only for companies located in Mexico and Brazil, where national markets would permit production of over 400,000 units for two. Even the combined markets of the five Andean countries (Colombia, Peru, Chile, Ecuador, and Bolivia) would yield a market of less than 200,000 vehicles by 1980 on the above assumptions. If they wished to produce at costs approximating international levels, the Andean countries could require only about 40 per cent local con-

---

* In technical terms, the average price elasticity found for automobiles in several advanced countries was about −1.5 while that for Latin America has been estimated at −1.2. Income elasticity averaged +2.5 for the advanced countries, but +1.9 for Mexico and +2.1 for Brazil, averaging +2.0 for the larger markets.

tent; Venezuela maybe 60 per cent. The balance-of-payments costs to each would be over $100 million annually. If Venezuela were to join the Andean group the total output could rise to over 500,000 vehicles, making a one-firm industry sufficiently efficient to match international prices.

Table A-2
Motor Vehicle Demand Projections, LAFTA Countries

	1969 Sales (units)	GNP Growth Rate 1960–67 (per cent)	Assumed Annual Growth of Sales (per cent)	Estimated Sales 1975 (units)	Estimated Sales 1980 (units)
Brazil	353,000	3.9	7.8	555,000	800,000 to 950,000
Mexico	163,000	6.6	13.2	343,000	640,000 to 850,000
Argentina	218,000	2.8	5.6	260,000	394,000 to 580,000
Colombia	20,000	4.3	8.6	32,800	49,000 to 82,000
Venezuela	62,000	5.0	10.0	110,000	178,000 to 223,000
Chile	22,000	4.1	8.2	34,000	53,000 to 82,000
Peru	25,000	5.7	11.2	47,000	84,000 to 115,000
Other LAFTA	15,000	4.5	9.0	25,000	39,000 to 59,000
Total	867,000			1,388,000	2,202,000 2,899,000

The choices facing countries concerning integration are those relating to content requirements (national versus regional and how much?), volumes of production (national and regional), the extent and nature of specialization (should it be by concentrating all production under one or two national companies or by specializing in parts production to serve half a dozen major companies operating through the region?), and the structure of companies within a regionally integrated automobile industry.

Appendix B
The Petrochemical Industry

The Latin American petrochemical industry has had its major growth within the last ten years or so. This growth has resulted largely from import substitution and has depended on imported technology. Prior chemical production was based on nonpetroleum raw materials and used processes which were technically less advanced. To obtain the advanced technology in a short time, these countries also accepted foreign ownership of much of the petrochemical industry. The most rapid development has been in the countries having the basic raw materials—Mexico, Brazil, and Argentina. Though Venezuela has ample supplies of petroleum, it joins Colombia, Peru, and Chile in remaining so industrially underdeveloped that the internal market has been inadequate to support efficient production, and their costs are too high to permit selling in the world market.

Present Structure

As in the automotive industry the most important characteristics of the petrochemical development for the standpoint of integration are the structure and conditions of production, the ownership patterns, the extent of national integration, trade patterns, and prices.

National Production

The production structure in petrochemicals is dictated by the types of inputs available (and their costs), the technology employed, and the market demand. Because of its relatively favorable position in each of these aspects, *Mexico* stands first in present capacity of petrochemicals in Latin America. In comparison with the advanced countries, however, it is not a major producer, and it supplies only about three-fourths of its own consumption of basic petrochemicals. Mexico is well endowed with the petrochemical building blocks of natural gas and petroleum, ranking fourth in the world in gas and twelfth in petroleum. However, it flares about one-fourth of its annual natural gas production. Mexico also is well-endowed with sulphur and phosphate, and the government-owned Petroleos Mexicanos (PEMEX) helps assure petrochemical producers of supplies of steam, electricity, fresh water, and technically trained manpower.

The author of this paper, Thomas S. Goho, is Assistant Professor of Economics at New Mexico State University.

149

Table B-1
Location of Production, Ownership, and Capacity of Selected Petrochemical Products (thousands of tons)

	Argentina	Brazil	Mexico	Venezuela	Colombia	Chile	Peru
Ethylene	Dow-13 Ko-15 ICI-15 Con-13 Dow-118[e]	UC-125[d] Sol-6 Ko-12 Ph-181[d]	G-27 G-27 G-36	G-150[e]	G-20 G-13[e]	G[a]-10[d]	
Ammonium Nitrate	Con-40	G-12 G-132[e] Ph-244[d]	G-99 L-66	G-55	G-34		Fo-2
Ammonium Sulfate	G-50	Ph-225[d]	G-263 L-88 L-45 L-46 L-67 L-11	G-85			Fo-20
Ammonia	AFP-68 G-1	Ph-180[d] RP-12 G-73 G-40	G-60 G-66 G-132[d] G-330[d] G-20 L-36	G-33 G[a]-216[e] CVN-150[e] G[a]-450[e]	G-126 G-10 CVN-150[e] IPC-90	L-1 ICI-1 G-330[e]	Fo-43 Fo-36
Urea	L-55	G-50	G-305 L-50	Fo-3 G-70 G-270[e] CVN-200[e]	G-10 CVN-200[e]	G-300[d]	

Table B-1 (continued)

	Argentina	Brazil	Mexico	Venezuela	Colombia	Chile	Peru
Styrene	Ko-15 Con-15	Fo-10 Ko[a]-60[d] Ko-16 G-13 G-12	G-60				
Benzene	Con-33 G-88	UC-20 G-30 Ph-84[d]	G-105	G-10	G-40[d]		
Phenol	Hk-10 L-8	L-7 BB-ε	L-5 L-5				
Caprolacton	Du-10		L-15 G[a]-300[d]		G-20[d] G[a]-16[d]		
Butadiene	Con-34	G-80 Ph-235[d]	G-55	G-n.a.			
Acetone	L-11	RP-4					
Vinyl Chloride	Mon-5 L-10	UC-70 Sol-12	G[a]-20			Dow-20[e]	
Methanol	BO-16 L-16	BO-40 L-20 Fo-17[e]	G-15				
Carbon Black	Ca-30 Con-12	Ph-17 G-24	Ph[b]-15	L-3 Fo-10[e]	Ph-13 Ca-9		

Table B-1 (continued)

	Argentina	Brazil	Mexico	Venezuela	Colombia	Chile	Peru
SBR Rubber	Con-37	G-40	G-44[a]				
Polyethylene	Dow-50[c] K-12 ICI-15	UC-19 Sol-10 L-10	G-34 G-18 G-58[d]	UC[a]-50[c]	Dow[a]-15	Dow[a]-20	
Polystyrene	L-2 Ko-6 Mo-8	Ko-9 L-8 BASF-1 Dow-40[c]	Ko[b]-10	Dow-2	Dow-5		

Note: There is more capacity than represented in the data here since not all companies are included; information remains partial as to companies producing and capacities.

[a] Joint venture of international company and government.
[b] Some local interests also in project.
[c] Planned.
[d] Under construction.

AFP—American foreign power.
BASF—BASF.
BB—Bunge Born.
Bo—Borden.
Ca—Cabot.
Ce—Celanese.
Con—Consortium of international companies.

Cona—Consortium of international companies plus local interests.
CVN—Joint venture between Colombian and Venezuelan governments.
Dow—Dow.
Du—du Pont.

Fo—Foreign but exact ownership uncertain.
G—Government.
Gr—Grace.
Hk—Hooker.
Ho—Hoechst.
ICI—ICI.
IPC—International Petroleum Company.

Ko—Koppers.
L—Local.
Mo—Monsanto.
Ph—Phillips.
RP—Rhône-Poulenc.
Sol—Solvay.
UC—Union Carbide.

As a consequence of governmental efforts, its abundant inputs, foreign technology, and a growing market, Mexico has had the highest rate of growth in petrochemical production in Latin America over the past decade. It has built capacity in some products far beyond its present levels of production. For example, production of ethyl alcohol meets all the present domestic needs, but facilities remain 75 per cent idle. There is even idle capacity in a 'few intermediate products that are also imported. For example, in 1967, domestic production of phthalic anhydride met only 68 per cent of domestic demand, measured by uses of the chemical in derivatives and end products, though operations were only at 52 per cent of capacity. The reason for the gap is that imports of many end-use products were less expensive than the domestically processed items and there was no domestic production of other end products using this intermediate. Similarly, Mexico had 40 per cent of its polyvinyl chloride facilities idle in 1967 but served only 60 per cent of the Mexican market. In polyethylene the same figures were 20 and 50 per cent; in styrene, 65 and 75 per cent. In fact, among ten major materials, intermediates, and final products, there was substantial idle capacity, whether or not the market was fully served; the other five were acetone, styrene, benzene, toulene, and vinyl acetate monomer. The Mexican government is stretching to build as complete a petrochemical complex as possible, inducing the building of facilities substantially beyond consumption levels in the hope of creating a strong nationally-integrated industry.

Brazil's drive toward the same goal is handicapped by the lack of large supplies of petroleum and natural gas; it ranks fifth in Latin America in petroleum and eighth out of ten that produce natural gas. Domestic petroleum output meets only about 30 per cent of national consumption; the shortage is supposed to be relieved by the long-term purchase of petroleum from Russia (rather than Venezuela, which has a supply many times its needs). Brazil also lacks sulphur for the production of sulphuric acid, which is a complementary input for much of the secondary industry that would be built on a petrochemical complex.

Brazil's major asset on the input side is availability of low-cost labor. Its wage rates are the lowest among the three largest countries, but higher than those within the Andean countries. The major asset needed in building a petrochemical complex is a large-size domestic market, which can sustain large-scale production facilities. This asset had not been fully used by Brazil as of 1968; consequently, Brazil's capacity in petrochemicals (not necessarily other chemicals) was lowest among the three major countries.

This situation is being remedied by the import of foreign capital and technology through Phillips Petroleum Company and Union Carbide. These two projects are to produce mainly ethylene and benzene, with output in 1975 to reach 675,000 tons, which would meet the country's total projected demand for these two chemicals in 1976. Brazil is also moving to shift over the technology used in production of a variety of chemicals that have been produced by nonpetrochemical routes; e.g., plants for butadiene and acetylene will be added to the other two. Several chemical products will be made by both petrochemical and nonpetrochemical techniques. Brazil plans by the mid-1970's to be largely self-sufficient in the production of basic and intermediate chemicals, produced via petrochemical methods. The items which Brazil produces and projects for the near future match those produced in Mexico: benzene, toulene, PVC, polyethylene, styrene, polystyrene, phenol, phthalic anhydride, vinyl acetate monomer, SBR rubber, etc.

As a consequence of the rapidly expanding market, Brazil is currently operating many plants at close to capacity and is still unable to meet demand. The 1967 styrene facilities were operating at 80 per cent of capacity, but still meeting only 50 per cent of the domestic demand; similarly in benzene. In mid-1969, 45 per cent of the Brazilian chemical producers reported that they were operating at capacity; only 17 per cent were not expanding their output because of stagnating demand. Brazil has not matched the growth in Mexico and has not as yet generated a large overcapacity. Overcapacity does exist in polyvinyl acetate and will occur elsewhere as the shift is made to modern, large-scale operations.

Argentina is relatively well endowed with material inputs for the petrochemical industry, ranking third in petroleum and fourth in natural gas. The slow pace of exploration and development of these resources had contributed to the slow rate of growth in the petrochemical sector. In addition to the petrochemical inputs, Argentina has an ample supply of resources needed for the development of complementary chemical products, but it does not have sufficient sulphur. From the standpoint of manpower, Argentina has the necessary technical training facilities, but its wage rates are third highest in Latin America, behind Venezuela and Chile.

Argentina started in the petrochemical industry before any other Latin American country, but Mexico quickly overtook it, so that by 1968, only 29 per cent of Latin American basic petrochemical capacity was located in Argentina. Like Brazil, the backbone of chemical production in Argentina remains nonpetrochemical technology, though Argentina is ahead of Brazil in using petrochemical methods.

Like the other major countries, Argentina is trying to obtain large-scale petrochemical capacity. It has turned to Dow to undertake establishment of a major petrochemical complex, producing the intermediates for hundreds of the more elaborate final products. The facility constitutes the largest single foreign investment in Argentina in a decade. It will produce ethylene, propylene, chlorine, and caustic soda. Despite these new building blocks for final products, Argentina will remain dependent on nonpetrochemical techniques for several major items, including acetylene, benzene, and ethyl alcohol. There is overcapacity in many of these items now, meaning that there is no pressure to change into more modern techniques through creation of new facilities. Only in ethylene and polyethylene was Argentina operating at more than 80 per cent capacity in 1967, while meeting full domestic needs. In fourteen other major products, capacity operations ranged from 40 to 80 per cent. Again, the list of products tracks with those in Mexico and Brazil: benzene, ethylene, butadiene, PVC, polyvinyl acetate, vinyl chloride, styrene and polystyrene, phenol, phthalic anhydride, SBR rubber, etc.

The western Latin American countries face quite similar production problems. Though they are not equally endowed with inputs, they all have one or both of the basic materials (petroleum and natural gas), and Venezuela is one of the largest petroleum producers of the world. The basic problem facing each and all of the Andean countries is an under-developed market for petrochemical products. Consequently, the combined output of the chemical industries in the four largest countries—Venezuela, Colombia, Peru, and Chile—is less than that of Argentina, the smallest of the three big countries. The chemical products now being made are based almost wholly on nonpetrochemical processes, using agricultural inputs, e.g., the production of ethyl alcohol via the fermentation of molasses.

In 1966 there were only five petrochemical plants in the western region: Colombia and Venezuela each had a carbon black plant for tires, and Colombia, Venezuela, and Peru were producing ammonia from natural gas for fertilizer. Major growth since then has been in the area of substitution for continued imports of fertilizers. Not only are facilities being duplicated within the region, but there is also considerable overcapacity among the products made by nonpetrochemical processes; capacity is utilized only about 50 per cent. There is, therefore, little pressure to shift to the more modern processes; even if there were, the items would be the same as those already being undertaken on large scales by the three major countries, which also need exports to utilize their own capacities. This duplication is evidenced by the recently reported arrangement between Dow and the Colombian government-owned company to produce a range of inter-mediates.

In sum, the three major countries are moving along the same road: to produce for the national market a wider range of intermediates and final products under more modern techniques. (The Andean countries, however, have hardly begun and therefore are far behind.) In every instance, the new developments are being duplicated and are highly dependent on foreign technology and assistance. This dependence has induced foreign ownership as well.

Ownership

In *Mexico*, the authority for developing and controlling petrochemical production is vested in PEMEX, the government-owned company having a monopoly on production of all basic chemicals. PEMEX is also charged with seeing that all secondary petrochemical producers are owned at least 60 per cent by Mexican interests, which may include itself. The list of items considered secondary was expanded to include products of many existing foreign-owned companies. The regulation has been applied to foreign-owned companies when they sought to expand operations, even though they were in existence before the rule was promulgated. For example, Union Carbide was producing polystyrene prior to the rule; thereafter, its output rose to 100 per cent of capacity and it needed to expand. The government insisted that the company sell 60 per cent to local interests as a condition of obtaining permission to expand. Rather than lose its control, Union Carbide sold the affiliate to another international company and a group of Mexican investors. Other preexisting operations have not yet been touched by the regulation; forty-five foreign-owned affiliates remain wholly owned.

Even though the role of PEMEX limits foreign-owned companies to producing some intermediates and final products, many U.S. and European chemical companies have sought entry into Mexico. Thirteen major U.S. chemical companies and five European have operations in Mexico; foreign capital represents 35 per cent of the total investment in the industry. The government will even subsidize a foreign company's entry if this is deemed desirable in order to cut imports or to complement an expansion by PEMEX. In exchange for this assistance, the affiliate must agree to sell its products at prices no higher than 15 per cent above those in the North American market, provided that raw materials are available at rates similar to those in that market.

The Mexican government would like to reduce foreign ownership but finds it cannot expand without the continuous inflow of technology. The international companies would like to retain control of their affiliates but will accept a minority position in order to be in the expanding Mexican market. The compromise creates a situation which reduces the interest of both the Mexican government and the international companies in integration with other countries.

In *Brazil,* the government-owned PETROBRAS has a monopoly in the production of petroleum and natural gas, but not in petrochemicals. The government has affirmed the right of private companies, either domestic or foreign, to undertake the development of the petrochemical (and chemical) industry. Many companies have responded, and most recently Union Carbide and Phillips have undertaken quite large projects in production of basic chemicals. Despite this encouragement to private investors, PETROBRAS is also in the field of basic chemicals, such as ethylene and benzene. It limits itself, however, to basic chemicals, staying out of the final products. Raw materials are produced by the state, but basic chemicals are produced by both state-enterprise and private companies.

The government has sought to encourage joint ventures, and many foreign companies have complied. For example, a Koppers Company affiliate is owned also by another international company (Huels A.G.), by PETROBRAS, and by private Brazilian investors. Koppers has provided the technology for a new styrene plant, with the raw materials to come from PETROBRAS; this joint venture is one of the few in which an American company holds a minority position. Compared to affiliates in Mexico, the Brazilian affiliates of American companies are more tightly under foreign control. During 1969 and 1970 a rash of interest has been evidenced by international chemical companies in Brazil, and many more joint ventures can be expected.

In *Argentina,* the government-owned YPF controls the exploration, development, and marketing of petroleum and natural gas, but it does not have a monopoly through ownership of facilities. Foreign enterprises are under contract to develop the resources and to market a set amount through YPF. The petrochemical industry itself is dominated by the international companies, since foreign enterprises have the same rights as domestic investors. These companies have invested in all phases of the industry; U.S. affiliates account for approximately 30 per cent of total sales in the industry, spread through the basic, intermediate, and final products. Most are wholly or majority owned. Of the twelve U.S. subsidiaries established between 1958–67, only one was held in the minority by the U.S. parent. Prior to that time, three were minority held, two others were majority held and at least six were wholly owned.

Apart from the production of chemicals for the military, the government has shunned direct ownership of chemical facilities and holds only a 10 per cent share of total chemical plant capacity. Even these government facilities rely on foreign technology. For example, Atanor S.A.M., a semi-government company, produced methanol under licenses from French and Italian companies. Argentine-owned companies also rely on foreign technology; for example, technology from one company combines Pechiney and Saint Gobain with some of its own developments to produce HCH. All international companies have brought technology with them into Argentina.

In *western Latin America*, each of the four major countries—Venezuela, Colombia, Chile, and Peru—have established state-owned companies for petrochemical production. Even so, little has been accomplished—in all but Venezuela—because of the absence of basic inputs. Assistance has had to be sought from the international companies. The major companies have been owned 100 per cent by foreign enterprises. Venezuela has established high long-range goals for development of the petrochemical industry based on use of domestic inputs; it hopes to build a complex which would export 80 per cent of the output of some plants to world markets. The project has bogged down, even with offers from the international companies to assist technologically, because of the inability of the government to devote sufficient resources to site development.

ECOPETROL, owned by the Colombian government, has sought joint ventures with international companies in oil production and refining and in the use of the by-products in intermediate products. It has teamed up with Dow in one substantial project; but in the main, petrochemicals are produced by 100 per cent-owned affiliates of the international companies: Grace, Dow, and Union Carbide. Monsanto holds a minority position in a company with local investors, and both Uniroyal and Goodrich produce tires (from petrochemical synthetics) in partnership with local capital.

Chile and Peru have also been dependent on foreign investment. Chile had to have foreign technology to produce polyethylene and polyvinyl chloride, and Dow responded with a joint venture with the government. Du Pont has held a wholly owned affiliate in Chile, and the German Farbwerke Hoechst has a joint venture. In Peru, Hoechst, and Farbenfabriken Bayer have contributed to petrochemical development.

Vertical Integration

Integration of the petrochemical companies back toward the raw materials has been prevented by the ownership and control of the early stages by the state-owned companies. In *Mexico* the typical pattern is for the local enterprises to buy from PEMEX intermediate monomers such as benzene and styrene. Private producers then process the styrene into polystyrene. Prior to 1966, affiliates of foreign companies imported the styrene from parent companies; by 1968, PEMEX's plant was on stream, and the source of styrene was shifted. This shift is an example of the nature of the integration problem within Mexican industry. While a type of national integration is occurring, no company is able to produce through all stages of output, from basic to final products. The effort of PEMEX is to untie foreign-owned affiliates from imports of intermediates and to substitute domestic sources. This move, coupled with the pressure to form joint ventures, makes it virtually impossible for the international companies to integrate their operations in Mexico with those in other countries and with the parent's sources of basic or intermediate products.

In *Brazil*, the international companies purchase some basic materials from PETROBRAS, but they are not prohibited from importing from the parent or of producing intermediates within Brazil. For example, Union Carbide obtains raw materials from PETROBRAS for production of ethylene, acetylene, and benzene; it combines acetylene with ethylene to produce vinyl chloride and can produce other end products as well. Such company integration *within* Brazil accelerates nationally oriented integration but discourages integration with the international companies' affiliates in other countries. Rapid internal growth should make further national integration attractive and reduce reliance on foreign enterprises as sources of intermediates and as markets for final products.

Argentina has the highest level of integration within companies, as a result of the absence of state ownership of intermediate stages of production. Major international companies have moved into the production of intermediates: PASA (a consortium of international companies), Koppers, ICI, and Dow. Others have begun plans to do the same. The scale of operations of the Dow project, only recently approved, will place Argentina on a par with many petrochemical complexes in North America and Western Europe. Argentina is seeking to achieve a nationally integrated industry, before moving to regional integration, by relying on international companies.

Western countries are so small marketwise that they cannot expect to achieve any significant degree of national integration. At present, neither Venezuela, Chile, nor Peru have the facilities for basic chemicals on which to build an integrated industry; for example, they are unable to take ethylene and convert it into a wide range of intermediate or end products. Colombia has a more developed industry, with Dow and Grace producing a fairly wide range of chemicals; but the industry will continue to be based on imported basic chemicals, until Dow's new project is completed.

Present efforts, therefore, among the Latin American governments are toward creating nationally integrated industries, relying heavily on international companies. Even in the western countries the possibility of starting with a *tabula rasa* and creating a regionally integrated industry is gradually being foreclosed.

Imports and Exports

One means of encouraging expansion of an integrated chemical industry is to shut out imports; but this tactic is likely also to restrict the market to the national economy, reducing the ability to achieve economies of scale. The Latin American countries have experienced just this result in the chemical industry. *Mexico* has rather low duties against chemical imports, but half of these, facing tariffs of 20 to 30 per cent, are permitted entry only under license. Licenses are granted only if the importer can prove that the high domestic price prevents him from exporting his final products to world markets. Between 1962 and 1968, imports decreased from 28 per cent of domestic consumption to 15 per cent while production more than doubled. But since consumption also rose, Mexico could not eliminate imports of basic chemicals.

Exports from Mexico amounted to just over 4 per cent of its chemical production in 1967; of this only 40 per cent were petrochemicals, such as aromatic chemicals and polyamide resins. The total petrochemicals amounted to about $22 million; less than one-third went to other Latin American countries; however, 80 per cent of the fertilizer exports went into intraregional trade. Affiliates of foreign-owned companies exported $12 million of petrochemicals, but only $1 million of this was exported to other affiliates or the parent company.

Although foreign-owned affiliates exported only 3 per cent of their production, they still did better than locally-owned companies in Mexico. The output is mostly end products with high value that could be transported at low cost, but the high prices tend to shut off export opportunities. European affiliates of the same international companies *have* been able to compete in world export markets.

To stimulate domestic production, *Brazil* applies a sliding scale tariff, rather than licenses. No duty is applied if the domestic industry is not capable of supplying more than a small percentage of the domestic needs of a given chemical. A duty of 20 to 40 per cent applies if domestic producers operating at or near capacity can supply between 50 and 80 per cent of the domestic consumption. A duty up to 55 per cent is applied if domestic production meets local trade needs, but if the industry is able to compete with imports without the duty, none is applied. There are exceptions to these rules of thumb, but they reflect the policy put into effect in April 1968.

A conflict arose between producers of plastics and of the intermediate monomers over interpretation of import policy. International companies were able to supply total Brazilian demands for the monomers and received the 55 per cent duty exemption, but the Plastics Industry Syndicate representing 1,500 local producers charged that the price of polyethylene and other monomers was twice as high as international prices, raising their costs and preventing them from selling end products in intraregional trade. The government resolved the issue in favor of the international companies. The tariff policy keeps imports of chemicals into Brazil quite low, amounting to only 16 per cent of consumption in 1962 and 1968. But in the petrochemicals group, Brazil has some way to go to achieve self-sufficiency, as indicated by the fact that the volume of these imports has risen from $150 million in 1964 to $270 million in 1968.

Exports of chemicals from Brazil have been virtually nonexistent, amounting to less than 2 per cent of total production, and have been largely nonpetrochemicals. Total chemical and plastics exports to other Latin American countries were roughly $20 million in 1965. The major petrochemical export was of SBR rubber during 1964–65 to other Latin American countries, amounting to $3 million the latter year. Afterwards, duty concessions made earlier under the national lists were terminated by the importing countries. By 1967, SBR rubber exports dropped to less than one-fourth of their 1965 level.

Of the total of $22 million of exports by the industry in 1967, Brazilian affiliates of U.S. companies reportedly accounted for approximately one-half, much larger than their 25 per cent share of total production. They accounted also for one-half of intraregional exports.

Argentina's tariffs have been the highest of the three major producers, not uncommonly exceeding 100 per cent. These rates often apply even when there is no local production; for example, the rate on sulphur is over 100 per cent, raising the price of sulphuric acid to three and four times the international level and, in turn, increasing the price of other chemical items using this input. A 150 per cent duty on caustic soda produces comparable results. Beginning in 1970, Argentina duties were cut to levels between 30 and 40 per cent and they are projected to be progressively lowered until they are between 10 and 20 per cent by 1976. The higher duties were undoubtedly effective in creating local production of a wide range of chemicals. Argentina, even with its smaller market, produces the same range as the two larger countries. Consequently, it has imported no more percentagewise than the others—about 16 per cent of total consumption—and the composition is similar to that of the other major countries. Still, several major chemicals, in the organic and plastics sectors, are not produced in the country.

As a result of high levels of self-sufficiency, Argentina has priced itself out of export markets; the industries promoted are still "infant." Argentina could export only 3 per cent of its total chemical production, with just over one-third going to other Latin American countries during 1967 and 1968. The bulk of these were nonpetrochemicals; glue alone was nearly one-third of the total, and quebracho (for tanning) was even larger. Argentine affiliates of U.S. companies reportedly accounted for 43 per cent of chemical exports but only 30 per cent of total production.

In the *western* countries, tariff rates range from 20 to 100 per cent on chemical products *not* produced within the countries, but this includes a range of 20 to 40 per cent in Colombia and between 80 and 100 per cent in Chile. Even higher duties are imposed on items which are produced within the countries; for example, rates on ethyl alcohol are 95 per cent in Colombia, 105 per cent in Venezuela, 206 per cent in Peru, and 453 per cent in Chile. Even so, local production by the petrochemical route has not increased rapidly because of the necessity to import substantial amounts of intermediates; for example, creation of a capacity in polystyrene requires imports of large amounts of styrene monomer, which is not feasible given pressures on balances of payments. Consequently, Peru and Venezuela import nearly 50 per cent of their total chemical needs, and Chile and Colombia between 30 and 35 per cent annually. The low level in Chile is not a result of larger local production but of the higher duties, which simply prohibit consumption.

These high rates and low efficiency make exports virtually impossible. Venezuela exports less than 1 per cent of its production and the others only around 2 per cent annually; less than 3 per cent of these exports are organic chemicals (those likely to be produced by petrochemical or more advanced technological routes). Almost all exports to other Latin American countries are of low-technology, high-labor content. Affiliates of U.S. enterprises have played a dominant role in the few petrochemical exports of the Andean region. It is estimated that three-fourths of Colombia's exports are composed of the exports of such affiliates to other affiliates in the same regions (Dow and Celanese).

In sum, all producing countries have pursued their nationally oriented policies through shutting out imports by high tariffs or prohibitions which have raised internal prices and thereby prevented exports of local production. The burden of exporting petrochemicals—small as these exports are —has fallen on affiliates of international companies.

Prices

Prices of petrochemical products are influenced in Latin America not only by high duties and import prohibitions but also by the fact of governmental monopoly in some countries in the production of basic chemicals. In Mexico, the prices PEMEX charges for its intermediates are reportedly not competitive with prices for similar items in the United States, being 30 to 40 per cent above international prices. One company's study of the petrochemical industry concluded that prices for nine major petrochemical products in Mexico ranged from 39 to 114 per cent above U.S. prices. The impact of such high prices is felt throughout the various stages of production: that benzene is 121 per cent above U.S. prices is a substantial reason for the high price of styrene (made from benzene), which is 43 per cent above U.S. levels.

Brazil's price structure is heavily influenced by the fact that it must import raw materials, which are not cheap to transport. Basic materials costs, plus high duties and high capital costs, are the major explanation of its high cost-structure: high-priced benzene leads to high-cost styrene and high-cost polystyrene. Despite its reliance on foreign sources of raw materials, compared to the Mexican state monopoly, Brazil does not come off much better. Its prices of fifteen major items compared to international levels range from 21 to 160 per cent higher; only SBR rubber has a price lower than international levels.

Argentina's cost structure is also influenced by high-cost basic chemical production. Unlike Brazil, it is largely self-sufficient in basic chemicals, but the plant size is substantially smaller than those in advanced countries. The resultant price structure is higher than that in either Mexico or Brazil. Compared to international levels, the prices of seventeen major items were 46 to 146 per cent higher.

Efforts to become competitive with international prices through large-scale production have been undercut by government policy. Several companies producing intermediates have erected plants of economic scale under official assurances, only to have the government permit another international company to enter, cutting the level of operations for both and leaving overcapacity. The overcapacity could be used only through regional integration or international specialization.

In summation, the following conditions obtain for all of the Latin American producers of petrochemicals, making regional (or international) integration both necessary and difficult at the same time:

1. Prices of products are relatively high compared to international prices, stemming from high costs of intermediates and basic chemicals.

2. The high prices of basic chemicals arise from high tariffs (even when there is little domestic production) and high costs of local production (usually because of small-scale production or state monopoly).

3. Substantial overcapacity exists in most of the basic-chemical production facilities, but prices are too high to induce others in Latin America to import from these sources.

4. Three distinct areas of production exist within the sector: basic chemicals, intermediates, and final products. The treatment of each differs somewhat within and among countries, but each is seeking to gain a high degree of *national* integration within the industry. Duplication of facilities is seen clearly in the data on selected products shown in Table B-1.

5. National integration is being achieved in Mexico through protection of final products, and in Brazil because of a rapidly expanding market. Integration is occurring very gradually in Argentina and has just begun in the western countries.

6. All countries remain dependent on international companies to achieve their desired levels of national integration, through the import of advanced technology; they have acceded to some levels of foreign ownership in the industry, some permitting vertical integration, others not. (See Table B-1.) But no country permits integration of the foreign-owned affiliates with the parent or other foreign affiliates, in or out of Latin America.

7. The inability to export high-priced items and the unwillingness to permit foreign-owned affiliates to integrate with the rest of the multinational enterprises owning them has meant that Latin America remains unintegrated with the rest of the world's chemical industry.

Recent pronouncements from government officials in Latin America assert that the region *must* obtain both a domestic petrochemical industry and one capable of surviving in international competition. To do this, the region must expand the markets open to each national industry—nationally, regionally, or internationally. The three types of market area call for three different policy approaches. The major countries seem to be following the first (expanding the national market); but Brazil appears to be the only one that may have a chance of success, and it is handicapped by the lack of domestic basic chemicals. The western countries seem to be moving simultaneously toward subregional integration and national rationalization. The alternative of regional integration through the removal of barriers within Latin America would produce substantial benefits and a first step was taken in December 1970 under the complementation agreement among Argentina, Brazil, Mexico, and Venezuela.

Gains From Integration

Given the small size of national markets, the national orientation of the petrochemical industry in each country, and the economics of scale which are obtainable in the industry, regional integration is considered by governmental officials to be an absolute necessity for the Latin American petrochemical industry if it is to become internationally competitive. Integration on any substantial scale would provide gains in the form of greater efficiency and reduced costs. The efficiency would come from a more continuous use of equipment and manpower; cost reduction would arise from this efficiency and from economies of scale. Economies from better utilization would depend on the coincidence of demand and production structure, which in turn depends on proper investment planning. For example, if a steam generating plant can be used in the production of several chemicals, its costs are spread over all of them. Economies of scale are predictable from experience in other countries, but the precise location of least-cost production in Latin America will be affected by the location of specific production factors and existing production facilities.

Economies of Scale

Economies are obtainable in three areas as scale is increased in the petrochemical industry: in investment outlays, in operating costs, and in cost of inputs.

The cost of plant and equipment in the petrochemical industry is a function of the surface area (external dimensions) of the equipment, not of the throughput. Since the surface area increases less rapidly than the volume of throughput, the investment cost tends to increase less than proportionally to output. A study for the United Nations covering eleven major products showed decreases in investment costs per unit of output ranging from 22 to 43 per cent as output was raised by a factor of three from a chosen base level.* The savings vary according to technologies used as well as the levels of output.

Normally, larger reductions in labor and supervisory costs occur as output is increased. For example, using the Udex method of producing benzene, plants producing from 20,000 to 100,000 tons per year may be constructed, with the number of workers per shift remaining at four. Production of ethylene by the cracking of ethane requires four workers in a 5,000-ton plant but only eight in a 100,000-ton plant. And, as the output of high-density polyethylene is increased from 8,000 tons to 48,000 tons, labor costs per ton drop by 50 per cent.

Similarly, supervisory cost remains the same over an even wider range of production, for the number of supervisors may remain the same whether only two workers or eight are needed to tend the critical points in the production process. For example, the production of phenol via the sulphonation of benzene requires one supervisor for four workers in a 3,000-ton plant, and the same supervisor can oversee the eighteen workers required in a 20,000-ton plant. The impact of the reductions in labor and supervisory costs cut per ton costs in ethylene by an additional 8 per cent over the 35 per cent drop arising from investment outlays as production moves from 10,000 to 60,000 tons per year.

Materials inputs, as well as fuel, electricity, steam and certain other services, tend to increase proportionally with output. They become more important relative to other costs as output increases, due to the reduction in other costs per ton of product. This increasing weight of materials input poses a dilemma for Latin American producers; the more they expand capacity the more important becomes the high cost of materials in determining their competitive position. The significance of new materials is enhanced by the fact that their supply and costs are relatively unstable in Latin America as compared to investment and labor costs; this makes the competitive position relatively unpredictable. Costs of new materials often are dependent on governmental policy, and producing companies are not permitted to integrate backwards to help make inputs more stable in price. Also, changes in government import policies on intermediates may change input cost substantially. However, governments could opt to integrate their petroleum and natural gas facilities regionally and stabilize supply and

* See Thomas Victoriez, "Programming Data for the Chemical Industry," *Industrialization and Productivity* (New York: United Nations, 1966), p. 55.

prices. If they did so, the costs could become competitive with international levels.

Variations in input costs are sometimes more important in affecting prices of final products than the cost reductions obtained through economies of scale. For example, the cost of ethylene in a plant using naphtha at a price of 7.5¢ per gallon can be cut from 2.4¢ to about 2.1¢ per pound when plant size is raised from 400 million pounds to 1 billion pounds. An increase in output by two and a half times cuts per unit cost by 12.5 per cent. But if naphtha is only 6.0¢ per gallon, it is cheaper per unit of output to have the smaller 400 million-pound plant. A 25 per cent increase in the feedstock price (from 6.0¢ to 7.5¢) has a greater impact on costs than does an increase in the plant size by two and a half times, because materials input rises to more than half of the cost.

Ethylene is itself a major input in many primary and intermediate products, and most petrochemical companies attempt to integrate backwards to assure themselves of stable and cheap supplies of this basic chemical. But in many countries of Latin America, such vertical integration is largely prohibited.

Costs are also substantially altered by the technology used in production, involving different inputs. For example, it is more expensive to produce ethylene in a 1 billion-pound plant with ethane at a cost of 3.0¢ per gallon than with propane at 4.5¢ per gallon. The added value of the coproducts from propane more than makes up the difference, assuming the ability to market the coproducts, which is dependent on demand.

The costs advantages from integration can be offset significantly by the costs of transportation within the region. The costs of transportation among the western countries alone are substantial, but the increased savings from integration would undoubtedly outweigh them.* However, among the three big producers—Mexico, Brazil, and Argentina—the scale of production is already such that, for many chemicals regional integration would bring less benefits in cost reduction than the increased costs of transport. Transport costs must be taken into account in any determination of the location of plants, if cost reductions are not to be dissipated.

Demand Projection

National markets in each producing country tend to be too small to permit the achievement of economies of scale; the smaller countries have had to limit this production to the end products where economies of large-scale production are not so important. Even the regional market for the Andean countries is not currently as large as that of any of the eastern countries.

---

* As shown in Chapter 6, "Transport Costs."

For example, the estimated demand for styrene in 1970 in the whole Andean group was 21,000 tons; Mexico's demand of 22,000 tons for the same item was the smallest among the big three.

Combination of the big-three markets at the present time would provide an adequate market large enough to induce efficient production for many basic and most intermediate chemicals, such as ethylene. And low-cost ethylene would reduce the cost of other products significantly. But combination of the three countries does not assure low-cost production in all major items, nor does present low-cost and competitive output guarantee future competitive position. Technology is changing so rapidly that what is a large-scale operation today may not be so tomorrow; a market that would support today's large-scale output may not be large enough to support tomorrow's technology. For example, several years ago, ethylene plants of 100 million pounds a year were common. To get a competitive lead, producers brought 300 million-pound plants on stream; today these producers find their competitive advantage undercut by plants with 600 million-pound capacities. There is little indication that the trend toward larger plants is abating, and an insufficiency of internal market demand—even with the whole of Latin America—seems likely for some products. This situation poses a dilemma in constructing Latin American plants: not to build them sufficiently large is to keep costs high, but to build capacity for the future is to build in obsolescence as technology changes.

Least-Cost Location of Production

Another dilemma is faced in Latin America. The greatest gains from integration will occur in the medium-sized countries, which have some production but not of efficient scale; but the lowest costs will tend to exist in the largest countries simply because the low-cost, medium-sized countries cannot produce *all* the chemicals in sufficient scale for the entire market.

Cost reductions in the medium-sized countries are illustrated by a typical sodium carbonate plant, with a present capacity of 54,000 tons per year; under integration it could be expanded to 360,000 tons, reducing cost by $20 per ton. In the large countries, plants of 216,000 tons already exist, and a shift to 360,000 tons brings a maximum saving of $5 per ton (and as low as $1 per ton depending on the processes used). The cost reductions arising in the intermediate products are substantially larger than those in basic chemicals, reflecting the higher unit value of the intermediates. For example, the cost reductions available to the medium-sized countries from integration in the intermediate products PVC and styrene would amount to $100 per ton, but only $30 for the large countries; the reduction available for methanol is only $10 in the medium-sized countries and zero in the large countries; that for butadiene is $20 in the medium-sized countries and zero in the larger.

At the other end of the line, the end products do not benefit from a larger scale of production to the same degree as the intermediates. Integration of plants producing SBR rubber, for example, would provide a $30 reduction in the medium-sized countries and $10 in the larger countries. Of course, where there are existing large-scale plants the gains are much smaller; in formaldehyde, which is produced in many countries on substantial scale, the cost gains are $5 in each set of countries. But in nylon 6, which is produced in few countries on any substantial scale, the gains are $125 in the medium-sized countries and between zero and $25 in the larger countries.

The location of the least-cost production under integration depends not only on the total size of the market but also on the resource structure of the individual country and the transport costs of getting materials and delivering products. Abstracting from transport costs of materials or products, the existing distribution of resources is such that Colombia tends to be the least-cost producer in Latin America as a whole for the large majority of important products. Among nineteen petrochemicals, Colombia was found to be potentially the least-cost producer of fifteen, sharing this position in one instance with Argentina and in another with Peru, while Venezuela is potentially the least-cost producer of the four remaining products.* But Mexico is potentially the second cheapest producer in practically every one of the nineteen items, sharing that position with Venezuela in only four items; Argentina was second in only one, tying with Mexico in another; and Brazil could only tie with Mexico for second place in one item. Potentially the least-competitive country, on a cost basis, is Chile; among the three largest countries, Argentina is potentially the least competitive.

The main causes of high costs in Argentina are the relatively high prices of electric power; labor; and inputs such as sugar, phosphate rock, and calcite (in other inputs it has relatively favorable prices). Brazil's cost disadvantage arises from the highest-priced petrochemical raw materials in Latin America, save for Chile. Mexico's disadvantage compared to Colombia arises from its high labor costs, which are also higher than those in Brazil and Peru; its electric power costs are also high. Venezuela is favored with the lowest prices for petrochemical raw materials, offset by labor costs seven times those of Peru and high prices for other materials inputs. Co-

---

* Colombia: ammonia, calcium carbide, benzene, butadiene, urea, vinyl acetate, polyvinyl chloride, styrene, cellulose acetate, nylon 6-6, and SBR rubber.
Venezuela: acetaldehyde, ethylene, acetic acid, and polyethylene (HD).
Colombia and Argentina: methanol.
Colombia and Peru: ammonium sulphate.
Source: United Nations Economic Commission for Latin America, *La Industria Química en America Latina* (New York: 1966), pp. 105–106.

lombia possesses the most favorable cost structure across the board because of its low labor costs and access to Venezuela's petroleum and natural gas; it holds this position not only in the specific chemicals indicated but also in broad groups of chemicals.

A study by ECLA calculated that Colombia would possess the lowest costs in an integrated petrochemical industry for each of the major product groups: primary mineral chemicals, organic chemicals, agricultural chemicals, plasticizers and synthetic resins, artificial fibers, and synthetic rubbers. And in only two of the major subgroupings was it displaced—by Peru in some acids and by Brazil in salts. Again, Mexico ranked a close second in all groups and major subgroups.*

Among the countries producing in the middle of the cost range, some differences exist among the product groupings, reflecting the resource endowment of each. For example, Venezuela makes up for its high labor costs with low-cost raw materials in products where labor input is relatively low, such as organic chemicals. Chile is in a strong position in basic acids because of its low-cost nitric and cupric raw materials; Peru has a similar situation. Argentina's comparative advantage rests in the production of olefins and industrial alcohol, due to its relatively low-priced natural gas.

Colombia is not endowed with a large enough supply of materials inputs, labor, and capital to produce *all* petrochemicals in the volume needed for Latin America; it would share output, at least with Mexico. But the rest of the countries would be wholly unwilling to let the industry settle in these two countries. And even those with a comparative advantage in a few items would likely be unwilling to accept that product or group as their only part in the industry.

Alternative Patterns for Integration

One method for assessing the trade-offs facing Latin American governments in approving integration for the petrochemical industry is to attempt to determine what the structure of the industry would look like under free trade within the region, based on an examination of relative cost factors.** Several limiting assumptions have to be made in order to approximate the result: that the transport system is efficient but not costless; that exchange rates are stable and expected to be so; that price levels are relatively stable; that the region will have sufficient labor, capital, and technology and will not

---

* United Nations Economic Commission for Latin America, *La Industria Química en America Latina* (New York: 1966), pp. 98–99.

** The data on which this "comparative statics" analysis is based were developed by the United Nations Economic Commission for Latin America, *La Industria Química en America Latina* (New York: 1966); and Latin American Free Trade Association, *La Industria Petro-Química en La ALALC* (Montevideo, April 1969).

have to alter the present ownership structure of the industry in order to obtain foreign inputs; that the special interests of state-owned enterprises are not permitted to interfere; that the Latin American market is the only one to be served; and that the objective is maximum *efficiency* in production of a selected range of products rather than in *all* petrochemical products. This last assumption means that it will be preferable to have lower costs for fewer products than more products at a higher cost.

The benefits of integration are calculated for two different patterns: integration among the three major countries and among the western (Andean) countries. The following sections indicate the structure of production that might be expected in each case and assess the possible benefits in terms of economic efficiency, balance-of-payments impacts, employment, and technology.

Major Countries

Projected Structure. Given the assumptions made above, the most efficient production will be located according to the availability of resources within each of the three countries. Mexico is the least-cost producer of most of the basic petrochemicals—notably ethylene, benzene, and orthoxylene. Mexico's cost of production of ethylene is $20 per ton less than in either Brazil or Argentina. Since ethylene is costly to transport (even with an efficient transport system), an agglomeration of intermediates production around the ethylene plant is typical even in advanced countries. Mexico would be expected to be the locus of production of a number of other chemicals, based on its low-cost ethylene; e.g., polythylene, which is projected to be $5 to $10 cheaper to produce there than in the other two countries if the three were integrated.

The low-cost production of ethylene and benzene makes Mexico the most efficient producer of styrene also; polystyrene follows naturally. But polystyrene is in such large demand in each of the countries that any one of them can sustain a plant of economic size serving its own market; in addition, if there is a need to transport, polystyrene is more cheaply produced nearer the market than nearer the production of styrene. If one works backward, the demand in each of the major countries is sufficient to support large-scale production of intermediate monomers, relying on basic materials provided by Mexico. But the cost of transportation of basic chemicals in bulk (having low unit-value) makes this arrangement costly. A trade-off arises for the three countries between high transport costs or high production costs of smaller-scale operations in basic chemicals.

Argentina's advantage lies in its supply of low-cost natural gas, which would normally lead to very efficient production of fertilizers. But Argentina's agricultural sector is not large enough to support large-scale production of fertilizers, and they are costly to transport in final form (given their low unit-value). It would appear most economic for Argentina to produce the ammonia for fertilizer, shipping it to Brazil, which would process it into ammonium nitrate and ammonium sulphate. Brazil has few of the raw materials necessary for fertilizer. Mexico appears capable of producing its own fertilizers since its natural gas is competitive with Argentina's, its local demand is adequate, and the high transport cost of the distance from Argentina to Mexico makes exportation of raw materials uneconomic.

Argentina has advantages in other intermediate products, however, arising from its supply of natural gas. Methanol is produced from ammonia, which is used in turn to produce methyl metacrylates; ammonia is also used in production of melamine and chlorofluoromethanes. But the total market of the three countries is not projected to be large enough for some time to support economic production of any of these intermediates. Therefore, what is left for Argentina is specialization in those items among the final products that are of high unit value and high technology—plastics, synthetic fibers, and rubbers.

Argentina is, in fact, the most technically advanced country in chemicals in Latin America, and it is logical that it should concentrate on that end of the spectrum. The problem of demand remains, however; most of the products, such as fiber glass, are consumed in high-income areas. Argentina would have to wait for economic development to raise income substantially before it could gain much from integration.

Brazil's advantage lies in production of chemicals related to heavy industry, for the demand is close by, largely within its highly developed automotive industry. The automobile and chemical industries are linked through the production and consumption of SBR rubber (tires and moldings), nylon cord, paint, vinyl, plastics, upholstery, etc. Economic scales of production in these end products is assured by the large-scale automobile output in Brazil. Besides this industry, Brazil has a large demand for chemically made rubber and could export into the world market through its traditional channels for natural rubber.

A third product group in which Brazil might specialize effectively consists of mass-produced, labor-intensive products, such as polyester fiber. It is the least-cost producer of this item, primarily because of its low-wage structure. As pointed out earlier, the fact that labor is a relatively small input in petrochemicals means that Brazil cannot capitalize on the low cost of this factor in determining its role. It is similar to Argentina in that it has an abundant resource, but the nature of the technology and structure of demand prevent it from being an adequate base for specialization.

Integration among the three major countries under free trade would likely result in a large concentration of petrochemical production within Mexico. This concentration is understandable when comparison is made with the location of production in the United States, where the bulk of production is in Texas and Louisiana, close to the relatively cheap sources of natural gas and petroleum.* Given the location of basic chemical production, a high degree of vertical integration within the same region arises to achieve the economies of scale at the level of intermediates. Decentralization arises at the next level of production, to the extent permitted by the cost of transport of intermediates, since it becomes more economic to produce final goods closer to the markets. For the major Latin American countries to move away from concentration of production in Mexico is to increase the costs of transport; transporting the basic chemicals would permit specialization at the level of intermediates. In the absence of free trade, or under a pattern of allocated production, higher costs of production of basic chemicals and intermediates must be accepted.

Potential Distribution of Benefits. Whether or not these countries will accept concentration of production will depend on the distribution of the benefits of integration among them. The areas in which benefits are counted as most important by governments include those of efficiency (and distribution of cost reductions), impacts on balance of payments, the balance of technology, and employment effects.

Efficiency in production—meaning costs of production compared to international levels—can be reached by integration among the three major countries in a large number of petrochemical products. Nearly the same gains in efficiency can be gained by integration among the three countries as can be gained by full Latin American integration. For example, costs of production of a group of chemicals for the *national* market in Brazil are about 134 per cent of what would occur under free trade over Latin America; in Argentina they are 146 per cent, and in Mexico 130 per cent. Three-country integration would bring the group down to 100 per cent of the free-trade cost in the full region; this level is equal to U.S. costs. Producers are carbon black, butadiene, ethylene, isopropyl alcohol, styrene, PVC, urea, polystyrene, methanol, polyethylene, and ammonia. Only a few of these products can be produced competitively on a purely national basis: carbon black and urea in Brazil and ammonia in Mexico. Apart from these, the cost differentials which would be eliminated by free trade in the three countries range from 7 to 95 per cent in Argentina, from 5 to 60 per cent in Brazil, and from 3 to 60 per cent in Mexico.

---

* Despite low labor-rates in the Southeast, little chemical production exists there, and what is there is relatively labor intensive. The Northeast is characterized by high incomes and technically oriented work force, but only a small share of basic chemical production exists there.

Given these cost reductions, integration would cut off imports from the United States and Europe of similar products and shift the pattern of trade. Mexico would eliminate some $40 million or 34 per cent of its annual chemical imports; Brazil some $30 million or 36 per cent; and Argentina some $20 million, or 30 per cent—based on the calculations of import substitution including only those items for which "minimally efficient" scales of production could be determined. The relatively lower shift in Argentina is accounted for by its previous efforts at import substitution and high tariffs, leaving it currently with less reliance on imports.

Trade within the region would also shift substantially. Based on estimates of demand during 1970 and the projected structure of industry, Mexico becomes the largest exporter. It would export some $40 million to Argentina and import less than $7 million from there; it would export some $57 million to Brazil, importing under $19 million. Mexico would gain an export surplus from the two others of $71 million. Trade between Argentina and Brazil would be approximately balanced at around $30 million each way, leaving both in deficit to Mexico at around $35 million. Taking into account the drop of imports from third countries, Argentina is still in deficit by $15 million and Brazil by roughly $5 million. This result reflects the fact that Mexico has already paid more attention to its export capabilities in developing its petrochemical industry, relying on its good endowment of resources.

In addition, Mexico's international position would be strengthened by integration, and it could expect to export even more to third countries. The international position of either Argentina or Brazil would also be strengthened, and they could expect to export to third countries items that they sold to each other or to Mexico. These additional exports would eliminate Brazil's deficit in the international accounts, and cut significantly into Argentina's. For example, Argentina would need to export only an additional 1 per cent of its production in 1970 to third countries in order to eliminate its deficit, but this is an increase of 30 per cent in exports; Brazil would need to sell only ⅓ of 1 per cent more of production abroad, or an increase of 10 per cent in exports. The experience in both the EFTA and EC indicate that integration in the industry does lead to increased exports to third countries, so this favorable result could be forthcoming.

Consequently, the trade impacts could well end up balanced for Argentina and Brazil, with Mexico gaining something more than $125 million in its accounts ($40 million of import substitution, $70 million net exports to Argentina and Brazil, plus expanded exports to third markets).

The third criterion of the appropriate distribution of the benefits of integration concerns the balance of technologies gained and used by each. Although it is difficult to compare the sophistication of two technological routes and to predict which should be used in what countries, it is clear from the shifts in production location that integration causes a shift in the locus of technically advanced production. Within their national orientations, the three countries have used basically the same technologies to produce the same items; even when a nonpetrochemical route was chosen—as in benzene, ethyl alcohol, and acetylene—all used the same technology.

Integration among the three would eliminate this parallel development, with Argentine and Mexican industries gaining the more advanced techniques. Mexico would become the center of large-scale production of basic and intermediate chemicals based on petrochemical processes; this shift to the new processes for basic chemicals and intermediates would significantly raise the level of technology in Mexico.

Argentina tends to be the center for the adaptation and dissemination of technology in petrochemicals within the subregion and would gain from the transfer of techniques it had tested and adapted from the international companies. Its specialization in deriving new chemicals and new uses would raise its technological level. But the means of raising its level is that of an expanded scientific and engineering base outside of manufacturing operations. Compared to Mexico, it is difficult to say which is the gainer.

Brazil, on the other hand, has natural and human resources more adapted to use of established processes, and does not have a large pool of skilled technicians and engineers to act as innovators or to disseminate or use the more advanced technologies on a large scale. On balance, it would gain the least technologically under free trade. Its advantage in producing the more sophisticated end-products for its large market will, eventually, let it move into the most advanced techniques in production of items such as butyl rubber, isoprene rubber, melamine, and ethanolomines. But this waits on the growth of a larger market.

Some of the gains in advanced technology must be balanced against the impacts of employment, which are also of concern to governments, both in the aggregate and within the particular industrial sector. Integration in petrochemicals would have greater impacts on employment outside of the industry than within. Integration would tend to drive out nonpetrochemical processes in production of several chemicals; for example, the labor-intensive process of manufacturing ethyl alcohol from molasses (and sugar cane) would be displaced. Losses in the labor-intensive suppliers of coal, limestone, and molasses would be recouped somewhat by increases in employment in the forwardly linked industries, using plastics, fibers, rubber, etc. It is impossible to strike a balance between these gains and losses without knowing the precise cases.

Given the low level of employment over wide ranges of outputs, it is unlikely that employment would be increased substantially in the manufacturing processes. In the EC experience, output increased between five and twelve times more rapidly than employment, depending on the country.

However, substantial employment would be created during the period of investment and restructuring of the industry among the three countries, if only by the construction of new facilities.

The distribution of gains and losses of employment depend on the relative position of the former suppliers in the chemical industry and the new, forwardly linked users of intermediates. Since each of the big three have been moving to petrochemical processes, and each still has a similar non-petrochemical base, similar losses in older industries would be faced, though the magnitudes will differ because of the size of the markets. Similarly, in the forward-user industries, size of the market would determine employment effects: Brazil probably would be the largest gainer because of its large consumer market. But Brazil's increases would probably be in the jobs of lesser skills, while those in Mexico and Argentina would lie in the more skilled tasks. The balance of gains and losses or of comparative benefits is difficult to cast up, and even precise knowledge of the numbers gained or lost would not provide a balance between skilled and unskilled employment nor meet each government's needs as to the location of employment changes.

Western Countries

A two-step move toward regional integration has been accepted within LAFTA, under which the Andean countries will integrate subregionally in preparation for full regional integration. The industrial structure would be altered twice, but since there is only a small base now, the question is the form of development. This section assumes that the market to be served is only the Andean group. Venezuela has been considered as a potential member of the Andean Pact, and is so discussed here; but its recent signing of a complementation agreement with the big three would seem to cast its lot outside of the Andean countries.

Projected Structure. Each of the four largest countries in the Andean region have one or both of the two petrochemical raw materials, while Ecuador and Bolivia have relatively less-developed resources. Their endowments are also quite varied in the other natural and human resources. Colombia and Venezuela are relatively rich in petroleum, while Chile and Peru have more abundant cupric, nitric, and phosphate resources. Colombia has relatively low-wages and low-incomes (per capita) compared to Venezuela; and Peru is a low-wage, low-income country compared to Chile.

These characteristics point to production of basic chemicals and intermediates in Colombia and Venezuela, with Venezuela the producer of ethylene at a price $35 to $45 per ton less than any other country in the region. All other basic and intermediate products are manufactured in Colombia at the least cost. Thus, Colombia and Venezuela would serve as the axis of chemical production in the Andean region. Venezuela, having the larger market, would undertake the high-income, market-oriented end products. Peru would be the logical location for fertilizer production since it provides the largest market for that item and possesses the nonpetroleum natural resources needed for production of ammonium nitrate, ammonium sulphate, and urea.

Under free trade within the subregion, Chile would be essentially left out of petrochemical production on several counts. It lacks the necessary natural resources; it is similar in other respects to the other three countries, adding no new demand dimension, and having natural resources similar to those of Peru. Even the high-income end products would tend to be produced in the larger Venezuelan market. Only excessive transport costs between Venezuela and Colombia and between these and Chile could make Chilean production economic under integration.

Just as Mexico would be the center of production under free trade in the eastern region, Colombia would acquire the complex of facilities producing basic and intermediate chemicals for the Andean group, exporting the latter to others for further processing. (This would be the result whether or not Venezuela joined the Andean group and is a prime reason why she opted to stay out.) Concentration would be even tighter than in the eastern region because the small total market demand does not allow for duplication of plants for any specific item.

Distribution of Benefits. As in the eastern region, the benefits under free trade tend to be skewed with the concentration of production but they vary with the levels of efficiency achieved, the shifts in trade patterns, the balance of technology, and employment impacts.

Efficiency levels cannot be as high as in the eastern region simply because the demand does not permit the scale of production in the Andean group to match even one national market among the big three. Before integration, the cost of production of a selected group of ten chemicals in Venezuela is 81 per cent above U.S. levels, in Chile 106 per cent higher, in Colombia 88 per cent higher, and in Peru 96 per cent higher. After subregional integration, average costs could drop to 20 per cent above U.S. levels, but in no product would the cost be likely to drop below those in the eastern subregion. However, costs of ammonia would drop to levels equal to those in the subregion. The distribution of the benefit of cost reductions would depend on the relative demand for specific chemicals, and the resulting patterns of trade.

The trade patterns of Colombia and Venezuela would be more affected by integration than those of Peru and Chile. The latter countries were not previously importers of many of the chemicals that would become available from the northern two partners after integration, so their imports would merely expand. Chile, because of its underdevelopment in chemicals, is likely not to have a significant demand for imports. Its major demand would be for fertilizers, produced in Peru, and some intermediates to support small-scale, end-product manufacture. It could hardly export these end products because of high labor costs. On balance, Chile would likely import new items amounting to $38 million, without exporting new volumes; it might eliminate some $4 million of imports from outside the region, but its net deficit would be nearly $34 million.

Peru would be able to export from its new fertilizer complex to a level of nearly $17 million, while importing about $33 million from its partners, for a negative balance of $16 million. This figure would be cut another $8 million by import substitution vis-à-vis countries outside the region, leaving it with a net deficit of $9 million.

Colombia and Venezuela gain the most from import substitution because they have been the primary importers of intermediates that would be produced within the subregion after integration. Colombia would be expected to eliminate some $17 million and Venezuela $13 million of former imports. They would gain some $120 million of the total $136 million of intraregional exports arising with integration. Therefore, Colombia would export $90 million, import $30 million, and substitute for former imports of $17 million, achieving a net favorable balance of over $76 million. Venezuela would export $30 million, import $35 million, and substitute for former imports of $13 million, for a favorable balance of $8 million. In the western region, as in the eastern, the payments benefits fall largely to one country—this time Colombia.

Colombia would also acquire the largest share of advanced technology, as a result of the concentration of production of basic and intermediate chemicals within its economy. Given the absence of permissible duplication, there is no spread of the same technology among the partners. The same small size of the market tends to limit the use of the most advanced technologies, relating to the larger scales of production. In any case, the locus of source-oriented chemicals would be Colombia and the market-oriented ones would be in Venezuela, leaving the others out of technological advances—save for Peru in fertilizers. Since Colombia and Venezuela are already the more-advanced countries technologically, integration would tend to accentuate the existing situation, rather than reversing it.

Employment in the western region also would follow the skewed distribution of the other benefits to Colombia, adding to both skilled and unskilled employment because of the locus of production of basic and intermediate products *and* the lowest wage level in the region. (Colombia plays the roles of both Mexico and Brazil in the eastern region.) However, the net increase in Colombia is mitigated by the fact that there would be a shift from labor-intensive, molasses-based production of ethyl alcohol to the petrochemical process.

Reductions in employment would not be suffered in Chile and Peru, even though they do not get substantial increases, because they are not currently significant suppliers of chemicals. Still, any reduction may be felt acutely, given the unemployment pressures in both countries that continue to exist.

In sum, Colombia is the net gainer from integration on all counts, with Venezuela following; neither Peru nor Chile would permit this distribution of benefits to occur, for it enriches the already relatively rich within the Andean group. Since Venezuela has refused to join, not wishing to take any chances in the development of its own petrochemical complex, it can be concluded that it did not like its second place or considered its potentials were greater with the big three.

Index

advanced countries, vs. less advanced, 3, 4, 22, 27, 28-31, 75, 149
agreements on surpluses and deficits, 4, 20
agricultural sector, 27, 155, 171
Alfa Romeo, 124, 129
Allende, Salvador, 126
Alliance for Progress, 104
allocation of production, *see* location of production
American Motors Corporation, 36, 128, 129
Andean Pact, 25, 26, 44, 71, 100, 101, 110, 114, 175
Andean subregion, 7, 15, 16, 17, 29, 39, 40, 82, 95
  automotive industry in, 50-53, 112, 114
  petrochemical industry in, 44-48, 153, 155, 157-158, 159, 161-162, 167, 175-178
antitrust activities, 13, 14, 98-99
Argentina, 7, 15, 16, 17, 20, 62, 89, 95, 96, 103, 104
  attitudes toward integration in, 111-112, 113
  automotive industry in, 28-29, 35, 36, 38, 40, 54, 56, 57, 59, 60, 74, 76, 111-112, 114, 117, 124-148 *passim*
  in bilateral trade, 28-29, 36, 74, 76, 111-112, 114, 136
  in complementation agreements, 23, 24, 25, 26
  ownership patterns in, 66-67, 128, 129, 130, 159
  petrochemical industry in, 30, 31, 41, 42, 43, 44, 45, 46, 49, 60, 61, 65, 67, 85, 114, 115, 149-175 *passim*
  pricing in, 136-137, 163
  trade patterns of, 21, 28-29, 50, 133, 134, 161, 173
Associated Spring Corporation, 130
Automex (Chrysler), 57, 129, 131
automotive assemblers, 36, 57, 58-59, 65, 72, 76, 85, 88, 94, 96, 112, 123, 126, 130, 131-132, 140, 141
automotive industry:
  benefits of integration in, 36-37, 139-148
  bilateral trade deals in, 28-29, 36, 74, 76, 111-112
  competition in, 36, 37, 39, 58, 81, 88, 114, 146
  and complementation agreements, 26, 28, 71, 80, 84-85, 86-87, 94-96, 110

local companies in, and integration, 113-114
local-content requirements in, 28, 35, 36, 39, 40, 87, 88, 112, 123-128, 133-137 *passim*, 144, 147
local-content vs. exchange costs in, 50-53, 74, 87
national lists in, 21
national policy in, 110-113
national production in, 123-128
nonscale cost differentials in, 143-146
ownership patterns in, 35-36, 123, 128-131
pricing in, 76-78, 136-139
production stages in, 140-143
scale factors in, 139-146 *passim*
specialization in, 37-39, 53, 54, 55, 56
trade patterns in, 28-30, 50-53, 58, 74, 75, 87, 123, 132-136
transportation in, 94-96
automotive parts and components, 11, 29, 36, 37, 38, 57, 76, 94, 95, 96, 125, 126, 140, 141, 144
  in intraregional trade, 132-136, 137
automotive stampings, 36, 72, 74, 125, 132, 133, 141, 142, 143
automotive suppliers, 36, 62, 65, 76, 77, 112, 137
  attitudes toward integration of, 114
  foreign ownership of, 129-130
  in joint ventures, 10-11, 57, 84
  local ownership of, 8, 9, 11, 36, 49, 57-60, 80, 123, 130
  in monopolies, 144
  and prices, 136-139
  in restructured industry, 84-85

balance of payments, 10, 11, 45, 76, 162
  complementation agreements and, 73-75, 97-98
  equity issue in, 1, 50-53, 173
  governmental interference and, 13, 14
Bendix Corporation, 130, 141
benzene, 44, 151, 153, 154, 155, 158, 159, 162, 165, 170, 174
Bolivia, 12, 55
  automotive industry and, 127, 147
  complementation agreements and, 23, 24, 70, 71
  petrochemical industry and, 175
Borgward, 131

179

About the Author

Jack N. Behrman taught at Davidson College, Washington and Lee University, and the University of Delaware in the fields of economics, business administration, and political science. In 1961 he joined the U.S. Department of Commerce and was appointed the following year to the new position of Assistant Secretary for Domestic and International Business. He left the government in 1964 to set up a program in international business at the School of Business Administration of the University of North Carolina. Among many special assignments, Behrman has been a consultant to the Pan American Union, the National Planning Association, the Organization of American States, and the U.S. Department of State. He has also been visiting professor at the Harvard Business School and lecturer at the Salzburg Seminar on American Studies, and he is a past president of the Association for Education in International Business. Behrman has written widely on the subjects of international finance, trade policy, and economic development. His most recent works are *National Interests and the Multinational Enterprise; U.S. International Business and Governments;* and *Multinational Production Consortia.*